Numerical Analysis

F. Owen, BSc(Econ.), AMBIM
and
R. Jones, BA(Econ.)

POLYTECH PUBLISHERS LTD STOCKPORT

First published in 1981

All rights reserved. No part of this publication may be reproduced, stored in a retrieval system, or transmitted in any form or by any means, electronic, mechanical, photocopying, recording or otherwise, without the prior permission of the copyright owner.

©Copyright Polytech Publishers Ltd.

Set in 10 pt Times series by
Bury Phototypesetting Limited, Peel Mills, Bury, Lancashire
and printed in Great Britain by
Ashworths Print Services, Peel Mills, Bury, Lancashire

Preface

This book has been specially written in response to the new syllabuses of the Association of Certified Accountants. It covers all the numerical analysis required for Part 1 of the examinations. It should also prove useful for other courses containing a mix of quantitative methods and statistics.

The approach used in this book is based on teaching similar courses for many years at Liverpool Polytechnic, an approach which reflects the growing belief that it is not enough merely to be able to perform calculations. Important as it is to have some degree of calculative ability, this alone is not sufficient. More and more it is becoming necessary for the student to understand, not only what he is doing, but also the meaning of the results he obtains.

We express our gratitude to all those who have made helpful comments and suggestions, but needless to say, responsibility for any remaining errors is ours alone.

As always, we are grateful to our wives for providing an environment conducive to the writing of this book.

F. Owen
R. H. Jones

Contents

Chapter		
1	Graphs and Gradients	1
2	Differentiation	9
3	Matrices	20
4	Compounding and Discounting	38
5	Investment Appraisal	52
6	The Organisation of Data	69
7	The Presentation of Data I	84
8	The Presentation of Data II	103
9	The Averaging of Data	123
10	The Time Series – I	143
11	The Time Series – II	152
12	Index Numbers	173
13	Dispersion	184
14	Statistical Sources	196
Answers to Exercises		212
Compounding Tables		217
Discounting Tables		220
Index		223

Chapter One

Graphs and Gradients

Functions

From a very early stage in human history man has been concerned with cause and effect. In fact, many of the most important scientific discoveries have taken place because someone has asked, "What will happen if I do this?". Much of mathematical knowledge is concerned with expressing the relationship between cause and effect in mathematical form. We know, for example, that the distance a body has fallen when dropped depends on the period of time for which it has been falling. The mathematician takes a general statement like this and makes it specific by saying

$$s = \tfrac{1}{2}gt^2$$

where s represents the distance fallen in time t seconds, and g is the gravitational constant, 32.

Now, an expression in this form extends our knowledge in two ways. Firstly, it tells us the exact way in which the distance fallen and the time are related; and secondly, it enables us to calculate the distance fallen if we know for how long the body has been falling. Naturally, if we had been given the distance fallen we could also calculate the time.

If you examine the above expression carefully you will notice one or two things about it. While s and t can take any values, the value of g is given as 32 and cannot change. Thus we may say that

> s and t are *variables*
> g is a *constant*

Looking now at the variables s and t we know that the value of s depends on the value of t, and so

> s is the *dependent variable* since its value depends on the value assigned to t
> t is the *independent variable* since it may take any value.

When two variables are connected in such a way that the value of one depends on the value of the other we say that the dependent variable is a *function* of the independent variable. Distance fallen is a function of the duration of the fall. Expressed more concisely we can put it that

$$s = f(t) \text{ (s is a function of t)}$$

while the expression $s = \tfrac{1}{2}gt^2$ tells us the precise form of the functional relationship.

Let us examine two such simple functional relationships that you might meet in business. Your transport manager is instructed to hire a car for the use of the company directors. He approaches a car hire firm who quote two different weekly tariffs.

Tariff 1. A charge of 7.5 pence per mile travelled.
Tariff 2. A fixed charge of £5 per week with an additional charge of 5 pence per mile travelled.

If we choose the first tariff we know that y (the cost) is a function of x (the distance travelled) and the precise form of the function is

$$y = 7.5x \text{ pence}$$
$$\text{or } y = £0.075x$$

We could draw up a table showing how the weekly charge varies according to mileage like this:

Mileage (x)	0	50	100	150	200	250	300	350
Cost £(y)	0	3.75	7.5	11.25	15	18.75	22.5	26.25

Looking at the table we can see that if 200 miles are covered in the week the charge is £15. We can plot this point on a graph.

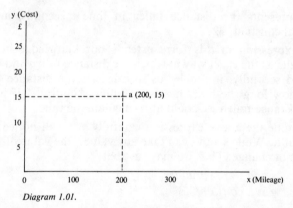

Diagram 1.01.

Having scaled the axes we locate 200 on the x (or horizontal) axis of the graph and imagine a line drawn vertically upwards from this point. This value of x is what we call the x *coordinate*. Now we locate 15 on the y (or vertical) axis and imagine a line drawn horizontally from this point. This value of y is called the y coordinate.

These two lines intersect at the point a, which can be identified by quoting its coordinates (200, 15). Any point on the graph can be identified if we know the coordinates. Thus the point (100, 7.5) is the point representing a value of x of 100 and a value of y corresponding to it of 7.5. If we do this for all the remaining points in our table, (0, 0), (50, 3.75), (100, 7.5) . . . etc., we find that it is possible to join all the points by a straight line, and we can read off the weekly charge (y) for any desired mileage (x) up to 350 miles.

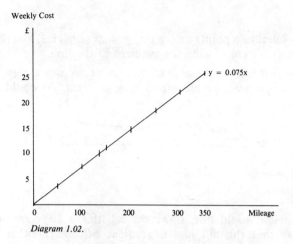
Diagram 1.02.

As we can read off the weekly charge from the graph we can say that the equation of the line we have drawn is the same as the equation describing the function, that is

$$y = 0.075x$$

Now the mileage charge may not be £0.075 per mile but some other value, say £m per mile. The equation of the line then becomes

$$y = mx$$

Any straight line which passes through the origin of the graph, i.e. the point (0, 0), has an equation of the form $y = mx$.

In the diagram below, three such straight lines have been drawn; $y = \frac{1}{2}x$, $y = x$, and $y = 2x$. Notice that the greater is the value of m, the steeper is the slope of the graph. In fact, m measures the *gradient* of the line.

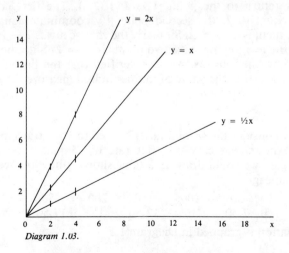
Diagram 1.03.

Gradients

We will now select two points on the line y = 2x, e.g. (1,2) and (2,4) and use them as a basis for calculating the gradient of the line.

Now what do we mean by a gradient? Suppose we saw a sign warning us that we were approaching a hill with a 10% gradient. We could represent it like this:

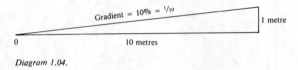

Diagram 1.04.

A mathematician would not, however, call this a 10% gradient; rather would he say that the hill had a gradient of $^1/_{10}$. In other words the mathematician calculates the gradient of a line by the formula

$$\frac{\text{vertical distance}}{\text{horizontal distance}}$$

Now if we consider a line drawn on a graph, then clearly the vertical distance is the change in the value of y and the horizontal distance is the change in the value of x. Hence we can state that the gradient of a line joining any two points on a graph is

$$\frac{\text{Change in y}}{\text{Change in x}} \text{ or as it is more commonly put } \frac{dy}{dx}$$

Let us now return to the point (1,2) and (2,4). The first point has a y coordinate equal to 2, the second point a y coordinate equal to 4. The change in y, then, is $4-2 = 2$. Similarly the change in x is $2 - 1 = 1$. So the gradient of the line joining these two points is $^2/_1 = 2$. Since both points lie on the straight line y = 2x we have verified that the line y = 2x has a gradient of 2 and that the value of m does indeed measure the gradient.

We will now consider the second tariff quoted to your transport manager, which you will remember was a flat rate of £5 plus 5 pence per mile travelled. Again we could draw up a table showing how the weekly charge varies with mileage.

Mileage (x)	0	50	100	150	200	250	300	350
Cost £(y)	0	7.50	10	12.50	15	17.50	20	22.50

This information is graphed in Diagram 1.5.

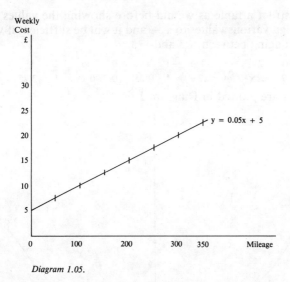

Diagram 1.05.

Again we can read off the weekly charge (y) for any mileage (x) up to 350 miles. For any mileage (x) the weekly charge is

$$y = £(0.05x + 5)$$

We can see that the gradient of this line is 0.05. Select any two points on the line and verify this for yourself.

Suppose that the hire charge was £c and the mileage charge £m per mile. The equation connecting weekly charge and mileage would then be

$$y = mx + c$$

Can you spot the difference between the line $y = mx$ and the line $y = mx + c$? The first line passes through the origin whereas the second line does not. What then can you conclude about the significance of c when drawing $y = mx + c$? Surely it is obvious that c determines the point at which the line cuts the y axis.

The gradient of a curve

The straight line has been useful in introducing concepts that we must master and thoroughly understand, and it is of far more use in the business world than many people think. But the functions that result in straight line graphs are only one type of function. The special characteristic of the *linear function* (as it is called) is that the equation contains only terms in which the power of x or y is unity, and constants. Suppose, however, that we are considering equations which contain terms such as x^2 or x^3, or those with terms such as $1/x$ or $2/x^2$. What sort of graph results then? Let us take as an example a situation where y is a function of x such that

$$y = x^2$$

We can construct a table as we did before showing the values taken by y when we assign various values to x — and it will be sufficient if we consider values of x ranging between +3 and −3.

x	−3	−2½	−2	−1½	−1	−½	0	½	1	1½	2	2½	3
y (= x²)	9	6.25	4	2.25	1	0.25	0	0.25	1	2.25	4	6.25	9

These figures are plotted in Diagram 1.6

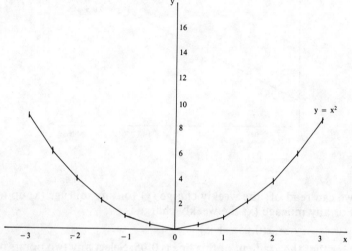

Diagram 1.06.

As you can see this graph is far from a straight line — it results in a U shaped curve which is symmetrical about the y axis. It is what we know as a *parabola*. The gradient of such a curve varies from point to point and we can no longer say that the gradient is measured by m. How then can we measure it?

Let us consider one small section of the parabola $y = x^2$, say between values for x of ½ and 1½. This section of the curve is drawn in diagram 1.7.

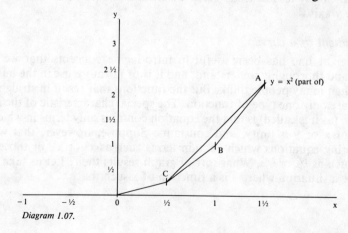

Diagram 1.07.

We will firstly consider the curve between point A (1½,2¼) and point C (½,¼). If we use the same technique as before we find that

$$dy = 2¼ - ¼ = 2 \quad dx = 1½ - ½ = 1$$

and we obtain a gradient $\frac{dy}{dx} = \frac{2}{1} = 2$. But this is clearly not the gradient of the curve — it is the gradient of the straight line AC which is a very different thing.

Consider now the point B (1,1) and C (½,¼). The gradient of the straight line BC is $\frac{1-¼}{1-½} = \frac{¾}{½} = 1½$. This is still not the gradient of the curve, but it is obvious from the diagram that the gradient of this straight line is much closer to the gradient of the curve than was that of the line AC. In fact, the closer B approaches to A, the more closely does the gradient of the line AB approach that of the curve. If we can allow B to approach so closely to A that the distance between them is infinitely small, the gradient of the line AB will differ from that of the curve by so little that the difference can be ignored. What we have done in fact is to make dx infinitely small so that in effect we are calculating the gradient of the curve at point A rather than between two points. The problem is that we have no means of measuring infinitely small distances by any technique that we have met so far.

The handling of this problem and the calculation of the gradient of a curve at any given point is a matter for differential calculus which we will introduce in the next chapter.

Exercises to Chapter 1

1.1 You decide to have a Spanish holiday and convert your holiday savings into pesetas. The rate of exchange obtainable from your bank is 148 pesetas to the pound. Draw a graph which would enable you to convert pesetas into pounds for amounts up to £5.00. Read off the sterling value of 115 pesetas; 326 pesetas; and 428 pesetas. (answers to the nearest ½ penny). What is the equation of the line?

1.2 Draw the graph of the line $y = -2x + 4$. What do you notice? What is the significance of the sign of m in the equation $y = mx + c$?

1.3 What is the gradient of the line joining the points (3,45) and (7,77)? Can you find the equation of this line?
(Hint: the equation of the line is $y = mx + c$. You have already calculated m, and we know that when $x = 3$, $y = 45$)

1.4 Consider the functions
$$y = 35x + 150$$
$$y = 25x + 200$$
Graph both these functions and hence determine the value of x which satisfies both equations.

1.5 Draw the curve of $y = 3x^3 - 9x$ for values of x between $+2$ and -2. For what values of x is the value of y a maximum; a minimum?

1.6 Consider the sketch of the curve y = x². It is easy to see that the line AB has a gradient of 3. We could allow A to approach B and recalculate the gradient of AB. The results are summarised in the table below where x and y refer to point A, dx is x minus one and dy is y minus one.

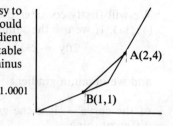

x	2	1.5	1.2	1.1	1.01	1.001	1.0001
y	4	2.25	1.44				
dx	1	0.5	0.2				
dy	3	1.25	0.44				
dy/dx	3	2.5	2.2				

Complete the table. What do you notice about the gradient?

1.7 If y = (x − 1) (x + 5) find the values of y when
x = 2, 1, 0, 9, (a + 1), −5

1.8 A motorist pays £50 road tax and £130 per year insurance. His car does 20 miles to the gallon and petrol costs £1.40 per gallon. His car is serviced every 3000 miles at a cost of £30. Depreciation per mile increases as mileage increases and can be calculated by multiplying the square of the mileage by 0.001 pence. If he does x miles per year derive an expression for his motoring cost per mile. Graph the expression you have derived and from the graph estimate the mileage that would minimise cost per mile.

Chapter Two

Differentiation

We will begin our study of differentiation by summarising what we have learned so far.

1. For any function, y = f(x), say y = $2x^2$, the value of that function will change as the value of x changes. If x = 2, the function has a value of 8; if x = 3, the function has a value of 18 and so on.
2. The rate at which the value of the function changes is of great importance in practice and we must learn how to calculate that rate of change.
3. If the function is linear, represented by a straight line graph, the rate of change is uniform, i.e. it is the same whatever the value of x, and we can measure it by measuring the gradient of the line. The gradient is equal to $\frac{dy}{dx}$ and this is constant along the line.
4. If the function has a graph which is a curve we have approximated the gradient of a curve by taking the gradient of the line joining any two points on the curve. Such a line is known as a *chord,* and we obtained better approximations to the gradient by taking successively shorter chords. In question 6 to the last chapter, we asked you to calculate the gradient of a number of such chords, each answer being a closer approximation to the actual rate of change of a function at a point. You have probably realised that if a chord joins two points on the curve which are infinitely close together, so that dx approximates to zero, the chord will be, in fact, the tangent to the curve. The gradient of the curve at any point will be the gradient of the tangent at that point.

In practice, although the gradient can be found by measuring the gradient of the tangent, it is not easy, and for practical purposes, (and for accuracy) an algebraic method is necessary.

Differentiation of x^2

Suppose we have a function y = x^2, any point on the resultant curve will have coordinates (x,x^2). Let us now choose a second point on the curve close to this first point such that the x coordinate is increased by a small amount to x + dx. The y coordinate will also be increased by a small amount to y + dy = $(x + dx)^2$.

$$y + dy = (x + dx)^2$$
$$= x^2 + 2xdx + (dx)^2$$

We can find dy by subtracting y from both sides.
$$dy = x^2 + 2xdx + (dx)^2 + (-y)$$
Since $y = x^2$
$$dy = x^2 + 2xdx + (dx)^2 - x^2$$
$$= 2xdx + (dx)^2$$

We can obtain the gradient $\frac{dy}{dx}$ by dividing both sides by dx

$$\frac{dy}{dx} = \frac{2xdx + (dx)^2}{dx}$$
$$= 2x + dx$$

Now if dx is very small, almost zero, the difference between $2x + dx$ and $2x$ is so small that we can ignore it, and we can say

$$\frac{dy}{dx} = 2x$$

What does this expression mean? Quite simply that if we take any point on the curve, the gradient of the curve at that point is equal to twice the x coordinate. When $x = 3$, that is at the point (3,9), the gradient is 6; when $x = 2$, the point (2,4), the rate of change of the function is 4. In dealing with the gradient of the curve in this way you must remember that $\frac{dy}{dx}$ is not a fraction, but merely a convenient way of indicating the gradient of a curve at a point.

Differentiation of $2x^3$

As before we will increase x by a small amount dx to $x + dx$. Since $y = 2x^3$,
$$y + dy = 2(x + dx)^3$$
$$= 2(x^3 + 3x^2(dx) + 3x(dx)^2 + (dx)^3)$$
$$= 2x^3 + 6x^2(dx) + 6x(dx)^2 + 2(dx)^3$$

Subtracting $y (= 2x^3)$ from both sides, we have
$$dy = 6x^2(dx) + 6x(dx)^2 + 2(dx)^3$$

Now dividing both sides by dx
$$\frac{dy}{dx} = 6x^2 + 6x(dx) + 2(dx)^2$$

If dx is very small $6x(dx)$ and $2(dx)^2$ will be so small they can be ignored and
$$\frac{dy}{dx} = 6x^2$$

A general rule

It would indeed be cumbersome if, every time we wished to differentiate an expression we had to calculate it from first principles in this way. Fortunately if we look at the two results we have we can derive a general method of differentiating.

If $y = x^2$ then $\frac{dy}{dx} = 2x$

if $y = 2x^3$ then $\frac{dy}{dx} = 6x^2$

If you look carefully at these two expressions you will see that in each case the coefficient of the *derivative* $\frac{dy}{dx}$ is equal to the coefficient of x in the original expression multiplied by the index of x in the original expression. For the first expression $1 \times 2 = 2$, and for the second, $2 \times 3 = 6$.
In each case too the index of the derivative is one less than the index of the original expression. So we can deduce the rule:

> To obtain the derivative of an expression multiply the index by the coefficient (giving the coefficient of the derivative,), and subtract one from the index, (giving the index of the derivative).

We can generalise this rule as follows:

> If we have a function $y = mx^n$
>
> The derivative $\frac{dy}{dx} = nmx^{n-1}$

Some aspects of differentiation

The above rule holds for all functions, not merely those in which the index is positive.

a) Suppose we are asked to differentiate $y = \frac{1}{x}$

In order to apply the rule we must put this expression into the form $y = mx^n$. We know that

$\frac{1}{x} = x^{-1}$ so our expression is $y = x^{-1}$

and $\frac{dy}{dx} = -x^{-2} = \frac{1}{x^2}$

b) In the same way if $y = \sqrt{x}$

$y = x^{½}$ and $\frac{dy}{dx} = ½x^{-½} = \frac{1}{2x^{½}} = \frac{1}{2\sqrt{x}}$

Thus, provided you can put the expression into the form $y = mx^n$ differentiation will present no problems.

c) Most expressions you will meet are rather more complex than the simple ones we have so far dealt with. A typical algebraic expression would be $y = 3x^4 + 2x^3 - 5x^2 + x - 8$

This more cumbersome expression is no more difficult to differentiate than the easier ones. All we have to do it to take each term of the expression in turn and differentiate that term on its own. The last two terms may however be a little confusing at first glance. Remember that

$x = 1x^1$ and hence $\frac{dy}{dx} = 1 \times 1x^0 = 1 \times 1 \times 1 = 1$

and $8 = 8x^0$ and hence $\frac{dy}{dx} = 0 \times 8x^{-1} = 0$

So if we differentiate $y = 3x^4 + 2x^3 - 5x^2 + x - 8$

we get $\frac{dy}{dx} = 12x^3 + 6x^2 - 10x + 1$

We find that in any expression, when we differentiate, the constants disappear. Thus when we differentiate 8 above we get the value 0, and this will be true no matter what the constant is.

d) So far we have differentiated only functions in which y is expressed as some function of x, and in which the differential is expressed as $\frac{dy}{dx}$.

You will find, however, that expressions take many forms; e.g. $s = \frac{1}{2}at^2$, where a is a constant. The correct notation for the derivative is, in fact,

$$\frac{d \text{ (the dependent variable)}}{d \text{ (the independent variable)}}$$

so if we were differentiating $s = \frac{1}{2}at^2$

we would say $\frac{ds}{dt} = at$.

In the same way, if we were differentiating $C = \frac{120}{x} + 600x$

we would say $\frac{dC}{dx} = \frac{-120}{x^2} + 600$

e) Sometimes the expression you are asked to differentiate will be the product of two or more other expressions. You might be faced, for example, with something like this:

Differentiate $y = (x^2 + 2)(x - 3)$

Now, there is nothing to stop you from multiplying out this expression and saying that

$y = x^3 - 3x^2 + 2x - 6$

and $\frac{dy}{dx} = 3x^2 - 6x + 2$

There will be times though when to multiply out is cumbersome, or even impossible; (for example $y = x^2\sqrt{1-x}$). We need, then, to have a rule for differentiating products similar to that which we have for differentiating a single expression. That rule is simply this:

> Differentiate one of the factors and multiply by the other. Now differentiate the other factor and multiply by the first. Add the two results and you get $\frac{dy}{dx}$.

Generalising, if we call the first factor u and the second factor v, the expression $y = (x^2 + 2)(x - 3)$ becomes $y = uv$

and
$$\frac{dy}{dx} = u\frac{dv}{dx} + v\frac{du}{dx}$$

Apply this rule to the expression
$$y = (x^2 + 2)(x - 3)$$
Let $\quad u = x^2 + 2$
and $\quad v = x - 3$
$$\frac{du}{dx} = 2x \qquad \frac{dv}{dx} = 1$$
$$\frac{dy}{dx} = (x^2 + 2)(1) + (x - 3)(2x)$$
$$= x^2 + 2 + 2x^2 - 6x$$
$$= 3x^2 - 6x + 2$$

which is precisely what we got previously when we multiplied out. This rule may be extended to as many factors as you wish. Suppose we had three factors, u, v and w and are asked to differentiate the expression $y = uvw$

$$\frac{dy}{dx} = vw\frac{du}{dx} + uw\frac{dv}{dx} + uv\frac{dw}{dx}$$

f) Rather similar to this, though perhaps a little more cumbersome, is the situation where the value of y is a quotient such as
$$y = \frac{3x}{x - 1} = \frac{u}{v}$$

In this case
$$\frac{dy}{dx} = \frac{v \cdot \frac{du}{dx} - u \cdot \frac{dv}{dx}}{v^2}$$

This can be stated in words:

> Multiply the denominator by the derivative of the numerator; subtract from it the numerator multiplied by the derivative of the denominator; divide the result by the denominator squared.

Thus if
$$y = \frac{3x}{x-1}$$
$$\frac{dy}{dx} = \frac{(x-1)(3) - (3x)(1)}{(x-1)^2} = \frac{-3}{(x-1)^2}$$

Why Differentiate?

Examine Diagram 2.1. This is the graph of the expression
$$y = 4x^3 + 3x^2 - 36x + 24$$
and, as you can see, the function is plotted between $x = +3$ and $x = -3$.

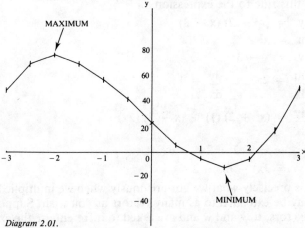

Diagram 2.01.

There is a maximum value of y when $x = -2$ and a minimum value of y when $x = +1.5$. Now there are many occasions when we want to know the value of x which makes the expression a maximum or a minimum. Suppose, for example, that we are building a container designed to carry a given volume of liquid. We can find many combinations of length, breadth and height which will give us this volume. But different dimensions to a container involve using differing amounts of material to produce it even though the cubic capacity remains the same. The surface area of a rectangular container, for example is given by:

2 (length × breadth + length × height + height × breadth)

Thus we can build a container of capacity 8 cubic centimetres by building it 8 centimetres long, 1 centimetre high and 1 centimetre wide. The material required would be

2(8 × 1 + 8 × 1 + 1 × 1) square centimetres = 34 square centimetres

We could, alternatively construct our box 4 centimetres long, 2 centimetres wide and 1 centimetre high. In this case the material required would be

2(4 × 2 + 4 × 1 + 2 × 1) square centimetres = 28 square centimetres.

Using the latter dimensions would reduce the material required by 17.6% – an appreciable saving.

The importance of differentiation is that it enables us to identify the value of x which maximises or minimises the value of an expression without having to draw the graph or engage in experiment as we have done above.

Maxima and Minima

Look again at diagram 2.1, this time concentrating on the gradient of the graph. While it is difficult to estimate what the gradient is at any point, we can see that at two points, (where the expression is a maximum and where it is a minimum) the gradient is zero. Now since the gradient is measured by $\frac{dy}{dx}$, we know that if we put $\frac{dy}{dx}$ equal to zero we have identified the values of x which make the gradient zero. We have found, that is, the values of x which make the value of the expression a maximum or a minimum.

Suppose
$$y = 4x^3 + 3x^2 - 36x + 24$$
$$\frac{dy}{dx} = 12x^2 + 6x - 36$$

Putting this derivative equal to zero we can find the values of x which satisfy the equation.

$$12x^2 + 6x - 36 = 0$$
$$2x^2 + x - 6 = 0$$
$$(2x - 3)(x + 2) = 0$$
If $(2x - 3) = 0$ $x = 1.5$
and if $(x + 2) = 0$ $x = -2$

There is a snag here, however. We know that when $x = 1.5$ and when $x = -2$ there is a maximum or a minimum value. But how can we tell which is which? After all, $x = 1.5$ could be either a maximum or a minimum value.

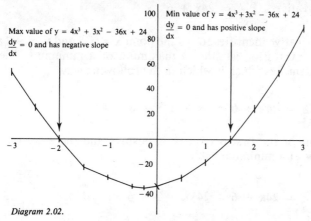

Diagram 2.02.

The key to discovering which it is lies in an understanding of the behaviour of the derivative. Diagram 2.02 shows the graph of the derivative

$$\frac{dy}{dx} = 12x^2 + 6x - 36$$

As you can see, it is a parabola. If we examine it closely we can see that the value of the derivative is zero when $x = 1.5$ and $x = -2$, which is precisely what we would expect. At these values the gradient of the original expression is zero. More important – when $x = -2$ and the expression is a maximum, the value of the gradient is falling. The curve is falling from left to right, i.e. it has a negative slope. When $x = 1.5$ and the expression is at a minimum value the derivative is increasing and the slope of the graph is positive. This relationship is always true. Whenever the value of the derivative is zero, if the slope of the graph of the derivative at that point is positive the original expression is at a minimum value; whenever it is negative the original expression is at a maximum value.

Now does this mean that to identify which is a maximum and which a minimum we have to draw a graph of the derivative? Of course not. You know that to find the slope of the graph of an expression we differentiate. Although the expression we are considering here is itself a derivative, there is nothing to stop us from differentiating this in its turn. We will then obtain what we know as the second derivative of the original function, indicated by $\frac{d^2y}{dx^2}$. Thus we obtain an expression, the second derivative, which will tell us, for any value of x, the slope of the first derivative.

We now have three expressions:

Original expression $\qquad y = 4x^3 + 3x^2 - 36x + 24$

First derivative $\qquad \frac{dy}{dx} = 12x^2 + 6x - 36$

Second derivative $\qquad \frac{d^2y}{dx^2} = 24x + 6$

The first derivative identifies for us the point $x = 1.5$ and $x = -2$ at which the original expression is either a maximum or a minimum. The second derivative identifies which is which in the following way.

At the point $x = 1.5$

$$\frac{d^2y}{dx^2} = 24x + 6 = 24 \times 1.5 + 6 = 42$$

The gradient of the first derivative is positive and therefore the original expression is at a minimum value.

At the point $x = -2$

$$\frac{d^2y}{dx^2} = 24x + 6 = 24 \times -2 + 6 = -42$$

For this value of x the gradient of the first derivative is negative and the expression is at a maximum value.

Example

M.A.T. Limited produce bottles of "jabra juice", a drink favoured by the witch doctors of West Africa. At present they are producing variable amounts each week according to the reports of their salesmen. Profits however are low and the managing director instructs his accountants to produce data showing his total revenue and his total costs of production. He then wishes them to determine that level of output which will maximise his firm's profits.

The accountants come up with the following information.

If q is the output of the firm per week in thousands of bottles

Total revenue is given by the expression $23q - \dfrac{q^2}{4}$

Total Cost is given by the expression $36 + 2q + 0.1q^2$

Since profit is total revenue minus total cost it follows that

Total Profit = $23q - \dfrac{q^2}{4} - 36 - 2q - \dfrac{q^2}{10}$

To find the value of q which will maximise total profit we must differentiate this expression.

Total Profit = $21q - \dfrac{14q^2}{40} - 36$

$\dfrac{d(Profit)}{dq} = 21 - \dfrac{28q}{40}$

For profit to be a maximum or a minimum

$\dfrac{d(Profit)}{dq}$ must equal zero i.e.

$21 - \dfrac{28q}{40} = 0$

$21 = \dfrac{28q}{40}$

$q = \dfrac{21 \times 40}{28}$ = 30 (thousand bottles per week)

But will this maximise or minimise profit? To find out we will take the second derivative.

$\dfrac{d^2(Profit)}{dq^2} = \dfrac{-28}{40}$

Consider the meaning of this carefully. The slope of the first derivative graph is independent of the value of q. In fact it is constant at $\dfrac{-28}{40}$.

Graphs with a constant gradient are, of course, straight lines, and this is what you would expect from the expression for the first derivative. It is of the form $y = mx + c$ where $c = 28$ and $m = \dfrac{-28}{40}$. More important from our point of view, the slope of the first derivative is negative, which means that the value of q which we have determined is a value which makes the expression for profits a maximum.

The accountant's response then would be:

"To maximise profits, produce 30,000 bottles of jabra juice each week."

Exercises to Chapter 2

2.1 Differentiate the following functions:

a) $y = 6x^2 + 5x - 7$

b) $y = \dfrac{5}{x} + 4x$

c) $y = 3x^3 + 3x - 1$

d) $y = \dfrac{4}{x^3} + \dfrac{3}{x^2} + \dfrac{2}{x}$

e) $y = (1 + x)^3$

f) $y = (8 - x)(x + 10)$

g) $y = (3x + 1)(2x + 1)$

h) $y = (ax^2 + bx + c)(px + q)$

i) $y = (x^2 + 1)(\tfrac{1}{2}x + 1)$

j) $y = \dfrac{x + 1}{x + 2}$

k) $y = \dfrac{2x^4}{x^2 - 4}$

l) $y = \dfrac{x^{1/2} + 2}{x^{1/2}}$

2.2 Determine which of the following has a maximum and which a minimum value:

a) $y = mx^2 - c$

b) $y = c - mx^2$

2.3 How many turning points do the following equations have?

a) $y = mx + c$

b) $y = ax^2 + bx + c$

c) $y = ax^3 + bx^2 + cx + d$

2.4 Explain fully each of the following:

a) A function $y = f(x)$

b) The first derivative of the function.

c) The second derivative of the function.

2.5 A motorist pays £50 per year road tax and £130 per year insurance. His car does 25 miles to the gallon and petrol costs £1.40 per gallon. His car is serviced every 6000 miles at a cost of £45. Depreciation increases as mileage increases and can be calculated by multiplying the square of the mileage by 0.0005 pence. If he does x miles per year find the mileage that would minimise cost per mile.

2.6 A manufacturer makes cylindrical containers of 400 cubic centimetres capacity. He wishes to minimise the amount of steel he uses. Find the internal dimensions of the can that would achieve his objective.

2.7 The total revenue of a manufacturer is given as $R = x(148 - x)$ where x is the quantity sold. Find
   a) the sales that would maximise total revenue
   b) the maximum total revenue
   c) the price he would have to charge to maximise total revenue.

2.8 A firm knows that its price (P) and its output (Q) are related by the expression $P = 80 - 2Q$
   a) Express Q in terms of P
   b) Use this expression to find an expression for total revenue in terms of Q
   c) Find the value of Q which maximises total revenue.

2.9 The costs of production of the same firm are composed of £125 in fixed costs, variable costs of £36 per unit produced and depreciation charges given by the expression £$0.05Q^2$. Find the level of output which would minimise average cost of production per unit.

2.10 The cost to a supplier of holding goods in stock depends on the amount he orders from the wholesaler. It is found that for John Smith Ltd the annual stockholding costs are

$$\text{Cost} = \frac{50 \times 2000}{q} + \frac{.001q}{4}$$

where q is the quantity ordered from the wholesaler. Find the size of order which will minimise the total cost of holding stock. If the supplier's customers demand 520,000 units a year, how often, on average, will the supplier place an order with his wholesaler.

2.11 A manufacturer knows that his total revenue is given by

$$\text{revenue} = 23Q - \frac{Q^2}{4}$$

His total cost of production is
$\text{Cost} = 36 + 2Q + 0.1Q^2$

where Q is the weekly production in thousands.

   a. Economists define marginal revenue as the rate of change of total revenue. Derive an expression for marginal revenue.
   b. How do you think economists might define "marginal cost"? Derive an expression for marginal cost.
   c. What output will make marginal revenue equal to marginal cost?
   d. Compare your answer with the output of "jabra juice" calculated in the last example in this chapter. What can you deduce?

Chapter Three

Matrices

In this chapter, we shall be examining a type of arithmetic that, in all probability, you will not have met before. Some of the concepts used will seem rather strange but their power and flexibility in application warrants their study. Moreover, computers rely heavily on matrices when performing complex caclulations or storing masses of data, so you will never fully appreciate computing without a sound understanding of matrix arithmetic.

What is a Matrix?

Let us suppose that a small firm is considering paying its workers a bonus on all output over a certain weekly level, and thinks that a bonus rate of 10p per unit would be appropriate. The firm wishes to know the weekly cost of this scheme, so the wages manager is instructed to undertake an investigation. He examines the production records of the workers concerned (suppose there are ten of them) and calculates the number of bonus units produced that week:

20   10   15   18   8   3   0   25   22   9

He has obtained what is called a *Set* of data, and it would be easy to calculate the bonus paid i.e.

$$0.1(20+10+15+....) = £130$$

Of course, we would have obtained the same result if the number of bonus units had been collected in a different order, i.e.

$$0.1(10+18+8+3+25+20+9+22+15+0) = £130$$

However, if the wages manager wishes to calculate the bonus each worker would receive, then the order of the bonus units would matter. To show that this is so, we would place brackets round the set.

(20   10   15   18   8   3   0   25   22   9)

The set has now become a *Matrix*. A matrix, then is a set of values where the position of the values is important. The implication of the ordering above is that the first worker receives 20 bonus units, the second worker receives 10 bonus units and so on. So if you see a set of numbers enclosed by brackets, always remember that the position of the numbers is important.

Now in this example, the matrix consists of a single row of numbers. It is a special kind of matrix called a *Row Vector*. Quite often, you will see the numbers written in a column like this:

$$\begin{pmatrix} 1 \\ 2 \\ 5 \end{pmatrix}$$

This type of matrix is called a *Column Vector*. If you think carefully, you will realise that if you fill in a football coupon, then you are constructing a column vector. Not only must the results be correctly forecast, but they must be forecast in the correct order — it is no use correctly predicting the total number of home wins, away wins and draws. For example, if you need to forecast the results of six matches, and use the symbols '0' for a draw, '1' for a home win and '2' for an away win then

$$\begin{pmatrix} 1 \\ 1 \\ 0 \\ 2 \\ 1 \\ 1 \end{pmatrix} \text{ and } \begin{pmatrix} 2 \\ 1 \\ 1 \\ 1 \\ 0 \\ 1 \end{pmatrix}$$

are both the same *set* of results, but are very different predictions. The *actual* results may form the same set i.e.

$$\begin{pmatrix} 2 \\ 0 \\ 1 \\ 1 \\ 1 \\ 1 \end{pmatrix}$$

but neither prediction would win. The sets are the same, but the vectors are different.

Now it would be perfectly possible to have a matrix with more than one row and more than one column. A single row or single column implies that the elements in the set are singly classified — in the case of the column vector above the elements of the matrix were classified according to the order of the matches. However, if a matrix has more than one row and more than one column, then a double classification is implied. For example, a firm may classify its labour force in two ways: by sex or by grade of work.

	Male	Female
Operative	400	100
Clerical	10	50
Executive	5	2

and the matrix

$$\begin{pmatrix} 400 & 100 \\ 10 & 50 \\ 5 & 2 \end{pmatrix}$$

describes the numbers of workers in each category. This matrix has three rows and two columns, and we would call it a (3 × 2) matrix. Moreover, if we use X to stand for any element in a matrix, then we can identify any particular element in the matrix. The value 5, for example is the element in the third row and the first column, so

$$x_{3,1} = 5$$

Likewise
$$x_{2,2} = 50$$
$$x_{2,1} = 10$$

Let us summarise what we have learnt so far. A matrix is a set where the position as well as the value of each element in the set is significant. If the matrix has a single row or a single column then it is called a vector. We state the dimensions of a matrix as (number of rows × number of columns) – *NEVER* the other way round. In fact this is the golden rule of matrices – everything is considered in a row by column ordering.

Matrix Addition and Subtraction

Suppose a plumbers merchant classifies the taps he holds in stock into bath taps, sink taps and garden taps. He also classifies his taps into imperial and metric sizes. The taps he currently holds in stock are:

	Metric	Imperial
Bath	105	32
Sink	220	45
Garden	96	32

Now the plumbers merchant thinks his stocks are dangerously low, and places an order with his supplier for the following quantities.

	Metric	Imperial
Bath	50	30
Sink	100	50
Garden	40	10

His new level of stock is:

	Metric	Imperial
Bath	155	62
Sink	320	95
Garden	136	42

Let us now suppose that the plumbers merchant receives an order from a plumber as follows:

	Metric	Imperial
Bath	10	5
Sink	20	5
Garden	10	2

The plumbers merchants level of stock is now:

	Metric	Imperial
Bath	145	57
Sink	300	90
Garden	126	40

We could have calculated the change in stock levels using matrices. When the builders merchant orders stock, stock levels increase so we apply matrix addition;

$$\begin{pmatrix} 105 & 32 \\ 220 & 45 \\ 96 & 32 \end{pmatrix} + \begin{pmatrix} 50 & 30 \\ 100 & 50 \\ 40 & 10 \end{pmatrix} = \begin{pmatrix} 155 & 62 \\ 320 & 95 \\ 136 & 42 \end{pmatrix}$$

When the plumber's merchant sells stock, stock levels decrease so we apply matrix subtraction.

$$\begin{pmatrix} 155 & 62 \\ 320 & 95 \\ 136 & 42 \end{pmatrix} - \begin{pmatrix} 10 & 5 \\ 20 & 5 \\ 10 & 2 \end{pmatrix} = \begin{pmatrix} 145 & 57 \\ 300 & 90 \\ 126 & 40 \end{pmatrix}$$

To add, (or subtract) matrices, add (or subtract) the corresponding elements. Addition and subtraction of matrices, then is a very simple process.

However, unlike ordinary arithmetic, not all addition or subtraction of matrices is possible. *It is only possible to add or subtract matrices of the same dimension.* For example, the following operation in matrices would not be possible.

$$\begin{pmatrix} 1 & 2 \\ 2 & 3 \\ 1 & 4 \end{pmatrix} + \begin{pmatrix} 1 & 3 \\ 2 & 2 \end{pmatrix}$$

although

$$\begin{pmatrix} 1 & 2 \\ 2 & 3 \\ 1 & 4 \end{pmatrix} + \begin{pmatrix} 1 & 3 \\ 2 & 2 \\ 0 & 0 \end{pmatrix}$$

is possible.

Matrix Multiplication

Now although addition or subtraction is a straightforward operation, matrix multiplication is not. In fact it is quite difficult to appreciate the logic behind matrix multiplication, and you would be well advised to master the technique first and worry about the logic later. There are two rules of matrix multiplication that you should always follow.

Rule 1.
Take in turn the rows of the first matrix with the columns of the second.
Rule 2.
Find the sum of the products of the corresponding elements.

Let us apply these rules to vectors.

$$A = (2\ 1),\ B = \begin{pmatrix} 2 \\ 3 \end{pmatrix}, \text{ find } AB$$

Now the first matrix has just one row, and the second matrix has just one column, so applying rule 2

$$(2 \times 2) + (1 \times 3) = 7,$$

So $(2\ \ 1) \begin{pmatrix} 2 \\ 3 \end{pmatrix} = (7)$

The result of this multiplication, then, is a single element vector. Notice that we have arrowed the corresponding elements i.e. those elements that must be multiplied together.

Let us see what would happen when we multiply a vector and matrix:

$$A = (2 \quad 1), B = \begin{pmatrix} 2 & 1 \\ 3 & 1 \end{pmatrix}, \text{ find } AB$$

$$AB = (2 \quad 1) \begin{pmatrix} 2 & 1 \\ 3 & 1 \end{pmatrix}$$

In this case, the second matrix has two columns, so we must take the columns of the second matrix in turn —

taking the first column

$$(2 \quad 1) \begin{pmatrix} 2 & 1 \\ 3 & 1 \end{pmatrix}$$

$$(2 \times 2) + (1 \times 3) = 7$$

taking the second column

$$(2 \quad 1) \begin{pmatrix} 2 & 1 \\ 3 & 1 \end{pmatrix}$$

$$(2 \times 1) + (1 \times 1) = 3$$

So

$$(2 \quad 1) \begin{pmatrix} 2 & 1 \\ 3 & 1 \end{pmatrix} = (7 \quad 3)$$

Now you may wonder why the product matrix is a row vector and not a column vector. This results from our final rule of matrix multiplication.

Rule 3

The position of the elements in the product matrix is determined by rule 1: the row position is determined by which row was used in the first matrix and the column position by which column is used in the second matrix.

The product $(3 \times 1) + (1 \times 1) = 3$ used the first (and only) row of the first matrix and the second column of the second matrix, so it takes the position 'first row second column' in the product matrix.

Armed with these three rules, we can now multiply a matrix by a matrix.

$$A = \begin{pmatrix} 2 & 1 \\ 4 & 1 \end{pmatrix} \quad B = \begin{pmatrix} 2 & 1 \\ 3 & 1 \end{pmatrix}, \text{ find } AB$$

Taking first row with first column.

$$\begin{pmatrix} 2 & 1 \\ 4 & 1 \end{pmatrix}\begin{pmatrix} 2 & 1 \\ 3 & 1 \end{pmatrix}$$

$$(2 \times 1) + (1 \times 3) = 7$$

$$\begin{pmatrix} 2 & 1 \\ 4 & 1 \end{pmatrix}\begin{pmatrix} 2 & 1 \\ 3 & 1 \end{pmatrix} = \begin{pmatrix} 7 \end{pmatrix}$$

Taking first row with second column.

$$\begin{pmatrix} 2 & 1 \\ 4 & 1 \end{pmatrix} \begin{pmatrix} 2 & 1 \\ 3 & 1 \end{pmatrix}$$

$$(2 \times 1) + (1 \times 1) = 3$$

$$\begin{pmatrix} 2 & 1 \\ 4 & 1 \end{pmatrix} \begin{pmatrix} 2 & 1 \\ 3 & 1 \end{pmatrix} = \begin{pmatrix} 7 & 3 \\ & \end{pmatrix}$$

Taking second row with first column.

$$\begin{pmatrix} 2 & 1 \\ 4 & 1 \end{pmatrix} \begin{pmatrix} 2 & 1 \\ 3 & 1 \end{pmatrix}$$

$$(4 \times 2) + (1 \times 3) = 11$$

$$\begin{pmatrix} 2 & 1 \\ 4 & 1 \end{pmatrix} \begin{pmatrix} 2 & 1 \\ 3 & 1 \end{pmatrix} = \begin{pmatrix} 7 & 3 \\ 11 & \end{pmatrix}$$

Finally, taking second row with second column

$$\begin{pmatrix} 2 & 1 \\ 4 & 1 \end{pmatrix} \begin{pmatrix} 2 & 1 \\ 3 & 1 \end{pmatrix}$$

$$(4 \times 1) + (1 \times 1) = 5$$

$$\begin{pmatrix} 2 & 1 \\ 4 & 1 \end{pmatrix} \begin{pmatrix} 2 & 1 \\ 3 & 1 \end{pmatrix} = \begin{pmatrix} 7 & 3 \\ 11 & 5 \end{pmatrix}$$

We can extend the dimensions of the matrices and still use the same multiplication rules, but beware! Not all matrix multiplications are possible. For example

$$\begin{pmatrix} 1 & 2 & 1 & 1 \\ 3 & 1 & 4 & 2 \\ 5 & 1 & 6 & 3 \end{pmatrix} \begin{pmatrix} 1 & 0 \\ 2 & 1 \\ 1 & 0 \\ 0 & 1 \end{pmatrix}$$

would give a product matrix

$$\begin{pmatrix} 6 & 3 \\ 9 & 3 \\ 13 & 4 \end{pmatrix}$$

but

$$\begin{pmatrix} 1 & 3 & 5 \\ 2 & 1 & 1 \\ 1 & 4 & 6 \\ 1 & 2 & 3 \end{pmatrix} \begin{pmatrix} 1 & 0 \\ 2 & 1 \\ 1 & 0 \\ 0 & 1 \end{pmatrix}$$

has no product matrix. Try to multiply the matrices and you will see that the system of rules breaks down. Taking 'first row with first column' we have

$$(1 \times 1) + (3 \times 2) + (5 \times 1) + \ldots$$

but there are insufficient elements in the row of the first matrix to complete the procedure: the matrices are not compatible. There is a simple rule for deciding whether matrix multiplication is possible – can you deduce it?

There must be the same number of columns in the first matrix as there are rows in the second.

Properties of Matrix Multiplication

In the last section, we used the two matrices
$$A = \begin{pmatrix} 2 & 1 \\ 4 & 1 \end{pmatrix} \text{ and } B = \begin{pmatrix} 2 & 1 \\ 3 & 1 \end{pmatrix}$$
and we found that
$$AB = \begin{pmatrix} 2 & 1 \\ 4 & 1 \end{pmatrix}\begin{pmatrix} 2 & 1 \\ 3 & 1 \end{pmatrix} = \begin{pmatrix} 7 & 3 \\ 11 & 5 \end{pmatrix}$$
Suppose that we now reverse the order of the matrices
$$BA = \begin{pmatrix} 2 & 1 \\ 3 & 1 \end{pmatrix}\begin{pmatrix} 2 & 1 \\ 4 & 1 \end{pmatrix} = \begin{pmatrix} 8 & 3 \\ 10 & 4 \end{pmatrix}$$
We can see that matrix multiplication differs fundamentally from algebraic multiplication. In algebra (or ordinary arithmetic) the product AB is identical to the product BA (after all $5 \times 3 = 3 \times 5$) but with matrices the product AB is quite different from the product BA. When multiplying matrices, it is vital that they are put in the right order, and to avoid confusion, mathematicians use the expression 'pre-multiply' and 'post-multiply' – rather than just say multiply. For example, the product matrix AB can be expressed as either 'pre-multiply B by A' or 'post-multiply A by B'. If we have an operation (for example, multiplication) to perform on a number of variables, and it does not matter in which order the variables are placed, then the operation is called *Commutative*. Thus, matrix multiplication is non-commutative.

Are there any exceptions to the rule that matrix multiplication is non-commutative?
$$A = \begin{pmatrix} 2 & 1 \\ 3 & 1 \end{pmatrix} \quad I = \begin{pmatrix} 1 & 0 \\ 0 & 1 \end{pmatrix}$$
$$IA = \begin{pmatrix} 1 & 0 \\ 0 & 1 \end{pmatrix}\begin{pmatrix} 2 & 1 \\ 3 & 1 \end{pmatrix} = \begin{pmatrix} 2 & 1 \\ 3 & 1 \end{pmatrix}$$
$$AI = \begin{pmatrix} 2 & 1 \\ 3 & 1 \end{pmatrix}\begin{pmatrix} 1 & 0 \\ 0 & 1 \end{pmatrix} = \begin{pmatrix} 2 & 1 \\ 3 & 1 \end{pmatrix}$$

Here, then, we have an operation that is commutative. The matrix I is a very special matrix: if the matrix A is either pre-multiplied or post multiplied by I, then A remains unchanged. The matrix I, then, acts in an identical fashion to the number one in ordinary arithmetic: for this reason I is called the *Unit Matrix* or *Identity Matrix*. Here, we have shown the identity as a 2 × 2 matrix, but is can have different dimensions to this. For example

$$I = (1), \quad I = \begin{pmatrix} 1 & 0 \\ 0 & 1 \end{pmatrix}, \quad I = \begin{pmatrix} 1 & 0 & 0 \\ 0 & 1 & 0 \\ 0 & 0 & 1 \end{pmatrix}, \quad I = \begin{pmatrix} 1 & 0 & 0 & 0 \\ 0 & 1 & 0 & 0 \\ 0 & 0 & 1 & 0 \\ 0 & 0 & 0 & 1 \end{pmatrix}$$

Notice the common pattern in these examples. Notice also that the identity matrix must be square (i.e. have the same number of rows as it has columns).

Now let us consider the product matrix AA^{-1} where

$$A = \begin{pmatrix} 2 & 1 \\ 3 & 2 \end{pmatrix} \text{ and } A^{-1} = \begin{pmatrix} 2 & -1 \\ -3 & 2 \end{pmatrix}$$

$$AA^{-1} = \begin{pmatrix} 2 & 1 \\ 3 & 2 \end{pmatrix}\begin{pmatrix} 2 & -1 \\ -3 & 2 \end{pmatrix} = \begin{pmatrix} 1 & 0 \\ 0 & 1 \end{pmatrix}$$

$$A^{-1}A = \begin{pmatrix} 2 & -1 \\ -3 & 2 \end{pmatrix}\begin{pmatrix} 2 & 1 \\ 3 & 2 \end{pmatrix} = \begin{pmatrix} 1 & 0 \\ 0 & 1 \end{pmatrix}$$

Again, we have a commutative operation: in both cases the product matrix is the identity matrix I. If one of the matrices is called A, then the other, A^{-1}, is called the *Inverse* of A. Why is A^{-1} called the inverse? Well, if you remember that the identity matrix is the matrix equivalent of the number one, then a parallel in ordinary arithmetic is

$$5 \times \frac{1}{5} = 1 \quad \text{or} \quad \frac{2}{9} \times \frac{9}{2} = 1$$

In both cases, the second number is the first number 'inverted'. Also, if you remember that 5^{-1} is $1/5$, you will realise why the inverse of A is A^{-1}.

Now let us consider the three matrices

$$A = \begin{pmatrix} 2 & 1 \\ 3 & 2 \end{pmatrix} \quad B = \begin{pmatrix} 1 & 2 \\ 3 & 1 \end{pmatrix} \quad C = \begin{pmatrix} 2 & 3 \\ 3 & 1 \end{pmatrix}$$

If we wish to find the matrix (AB)C, then we must combine the normal rule of brackets (which specifies the order for an operation) with the non-commutative rule of matrices. First we must find AB, then post-multiply by C

$$AB = \begin{pmatrix} 2 & 1 \\ 3 & 2 \end{pmatrix}\begin{pmatrix} 1 & 2 \\ 3 & 1 \end{pmatrix} = \begin{pmatrix} 5 & 5 \\ 9 & 8 \end{pmatrix}$$

$$(AB)C = \begin{pmatrix} 5 & 5 \\ 9 & 8 \end{pmatrix}\begin{pmatrix} 2 & 3 \\ 3 & 1 \end{pmatrix} = \begin{pmatrix} 25 & 20 \\ 42 & 35 \end{pmatrix}$$

Now let us find A(BC) This time, we must find BC first, then pre-multiply by A

$$BC = \begin{pmatrix} 1 & 2 \\ 3 & 1 \end{pmatrix}\begin{pmatrix} 2 & 3 \\ 3 & 1 \end{pmatrix} = \begin{pmatrix} 8 & 5 \\ 9 & 10 \end{pmatrix}$$

$$A(BC) = \begin{pmatrix} 2 & 1 \\ 3 & 2 \end{pmatrix}\begin{pmatrix} 8 & 5 \\ 9 & 10 \end{pmatrix} = \begin{pmatrix} 25 & 20 \\ 42 & 35 \end{pmatrix}$$

We can conclude that A(BC) = (AB)C or the position of the brackets does not matter in matrix multiplication. Using the jargon, we say that matrix multiplication is *associative*.

Let us now summarise what we have learnt in this section. Generally, matrix multiplication is non-commutative i.e.

$$AB \neq BA$$
$$\text{except that } AI = IA = A$$
$$\text{and } AA^{-1} = A^{-1}A = I$$

Also, matrix multiplication is associative, i.e.

$$A(BC) = (AB)C$$

Row Transformations: Finding the Inverse

If we set up a matrix multiplication sum something like this

$$\begin{pmatrix} 1 & 0 \\ 0 & 1 \end{pmatrix} \begin{pmatrix} 2 & 1 \\ 3 & 2 \end{pmatrix} = \begin{pmatrix} 2 & 1 \\ 3 & 2 \end{pmatrix}$$

then we can alter the rows of two of these matrices and the product matrix will still be correct. We will be performing *row transformations* and there are two rules which govern what we can do.

Rule 1

We can multiply (or divide) any row in the first matrix by a constant as long as we do likewise to the corresponding row in the third matrix.

For example, dividing the top row by 2 gives

$$\begin{pmatrix} ½ & 0 \\ 0 & 1 \end{pmatrix} \begin{pmatrix} 2 & 1 \\ 3 & 2 \end{pmatrix} = \begin{pmatrix} 1 & ½ \\ 3 & 2 \end{pmatrix}$$

You should check that the product matrix is correct

Rule 2

We can add (or subtract) any multiple of one row to any other row in the first matrix as long as we do likewise to the product matrix.

For example, subtracting three times the top row from the bottom row gives.

$$\begin{pmatrix} ½ & 0 \\ -3/2 & 1 \end{pmatrix} \begin{pmatrix} 2 & 1 \\ 3 & 2 \end{pmatrix} = \begin{pmatrix} 1 & ½ \\ 0 & ½ \end{pmatrix}$$

Multiplying the bottom row by 2 gives

$$\begin{pmatrix} ½ & 0 \\ -3 & 2 \end{pmatrix} \begin{pmatrix} 2 & 1 \\ 3 & 2 \end{pmatrix} = \begin{pmatrix} 1 & ½ \\ 0 & 1 \end{pmatrix}$$

and subtracting half the bottom row from the top row gives

$$\begin{pmatrix} 2 & -1 \\ -3 & 2 \end{pmatrix} \begin{pmatrix} 2 & 1 \\ 3 & 2 \end{pmatrix} = \begin{pmatrix} 1 & 0 \\ 0 & 1 \end{pmatrix}$$

At each of these stages, you should check that the product matrix is correct.

So we see that the rules for row transformations are indeed correct. So what? Well, if you look carefully at what we have done, you will realise that we have not been performing row transformations haphazardly. Let us define.

$$A = \begin{pmatrix} 2 & 1 \\ 3 & 2 \end{pmatrix}$$

We started with

$$\begin{pmatrix} 1 & 0 \\ 0 & 1 \end{pmatrix} \begin{pmatrix} 2 & 1 \\ 3 & 2 \end{pmatrix} = \begin{pmatrix} 2 & 1 \\ 3 & 2 \end{pmatrix}$$

i.e. $IA = A$

Whereas we ended with

$$\begin{pmatrix} 2 & -1 \\ -3 & 2 \end{pmatrix} \begin{pmatrix} 2 & 1 \\ 3 & 2 \end{pmatrix} = \begin{pmatrix} 1 & 0 \\ 0 & 1 \end{pmatrix}$$

which must be

$$A^{-1}A = I$$

In other words, we have performed row transformations in such a manner as to find the inverse of A! By changing the A on the right hand side of the equals sign into I, we have sympathetically changed the I on the left hand side into the inverse of A.

Let us examine more closely the use of row transformations to find an inverse. You should notice that at no time did the second matrix change, and it seems a waste of effort to restate it after each transformation. To avoid this repetition we can use a *partitioned matrix*. Suppose we wished to find the inverse of $\begin{pmatrix} 2 & 5 \\ 3 & 5 \end{pmatrix}$ Instead of the initial format

$$\begin{pmatrix} 1 & 0 \\ 0 & 1 \end{pmatrix} \begin{pmatrix} 2 & 5 \\ 3 & 5 \end{pmatrix} = \begin{pmatrix} 2 & 5 \\ 3 & 5 \end{pmatrix}$$

we can use the partitioned matrix

$$\left(\begin{array}{cc|cc} 1 & 0 & 2 & 5 \\ 0 & 1 & 3 & 5 \end{array} \right)$$

Can you see that if we transform the right hand side of the partition into the identity matrix $\begin{pmatrix} 1 & 0 \\ 0 & 1 \end{pmatrix}$, then we must transform the left hand side into the required inverse?

$$\left(\begin{array}{cc|cc} 1 & 0 & ② & 5 \\ 0 & 1 & 3 & 5 \end{array} \right)$$

Firstly, we will change the two (ringed) into 1. To do this, divide the top row by 2.

$$\left(\begin{array}{cc|cc} ½ & 0 & 1 & 5/2 \\ 0 & 1 & ③ & 5 \end{array} \right)$$

Next, we change the three into zero by subtracting three times the top row from the bottom row

$$\begin{pmatrix} 1/2 & 0 & | & 1 & 5/2 \\ -3/2 & 1 & | & 0 & -5/2 \end{pmatrix}$$

Adding the bottom row to the top row changes $5/2$ to zero

$$\begin{pmatrix} -1 & 1 & | & 1 & 0 \\ -3/2 & 1 & | & 0 & -5/2 \end{pmatrix}$$

Finally, multiplying the bottom row by $-2/5$ changes the $-5/2$ into one

$$\begin{pmatrix} -1 & 1 & | & 1 & 0 \\ 3/5 & -2/5 & | & 0 & 1 \end{pmatrix}$$

So the inverse of $\begin{pmatrix} 2 & 5 \\ 3 & 5 \end{pmatrix}$ is $\begin{pmatrix} -1 & 1 \\ 3/5 & -2/5 \end{pmatrix}$ and we can check that this is correct by multiplying them together. The inverse is correct if the product matrix is the identity

$$\begin{pmatrix} 2 & 5 \\ 3 & 5 \end{pmatrix} \begin{pmatrix} -1 & 1 \\ 3/5 & -2/5 \end{pmatrix} = \begin{pmatrix} 1 & 0 \\ 0 & 1 \end{pmatrix}$$

If we wish to find the inverse of a 3×3 matrix, then we would use exactly the same method, but this time use a 3×3 identity. For example

$$\begin{pmatrix} 1 & 1 & 1 & | & 1 & 0 & 0 \\ 2 & 1 & 2 & | & 0 & 1 & 0 \\ 1 & 2 & 2 & | & 0 & 0 & 1 \end{pmatrix}$$

Firstly, we will change the First column into $\begin{pmatrix} 1 \\ 0 \\ 0 \end{pmatrix}$ To do this we will

Subtract twice the top row from the middle row and
Subtract the top row from the bottom row.

$$\begin{pmatrix} 1 & 1 & 1 & | & 1 & 0 & 0 \\ 0 & -1 & 0 & | & -2 & 1 & 0 \\ 0 & 1 & 1 & | & -1 & 0 & 1 \end{pmatrix}$$

Now we shall change the second column into $\begin{pmatrix} 0 \\ 1 \\ 0 \end{pmatrix}$ To do this we will

Add the middle row to the top row
Add the middle row to the bottom row and
multiply the middle row by -1

$$\begin{pmatrix} 1 & 0 & 1 & | & -1 & 1 & 0 \\ 0 & 1 & 0 & | & 2 & -1 & 0 \\ 0 & 0 & 1 & | & -3 & 1 & 1 \end{pmatrix}$$

Finally, the third column must be changed to $\begin{pmatrix} 0 \\ 0 \\ 1 \end{pmatrix}$. To do this we must

Subtract the bottom row from the top row

$$\left(\begin{array}{ccc|ccc} 1 & 0 & 0 & 2 & 0 & -1 \\ 0 & 1 & 0 & 2 & -1 & 0 \\ 0 & 0 & 1 & -3 & 1 & 1 \end{array} \right)$$

Before completing this section, you should note two important points. Firstly, only square matrices can have inverses (remember the identity must be square). Secondly, even if a matrix is square then it still may not have an inverse: you can show that this is true by trying to find the inverse of

$$\begin{pmatrix} 2 & 1 \\ 4 & 2 \end{pmatrix}$$

Determinants

Finding an inverse can be a cumbersome business. Fortunately there is another method, but at this stage it is only really practical to apply it to a 2 × 2 matrix. Let us examine the matrix

$$A = \begin{pmatrix} a & b \\ c & d \end{pmatrix}$$

The quantity $ad - bc$ is called the *determinant* $|d|$ of the matrix and

$$A^{-1} = \frac{1}{|d|} \begin{pmatrix} d & -b \\ -c & a \end{pmatrix}$$

Notice carefully how the matrix A has been transposed in this expression. We shall now use this method to recalculate the inverse of $A = \begin{pmatrix} 2 & 5 \\ 3 & 5 \end{pmatrix}$

$$|d| = (2 \times 5) - (3 \times 5) = -5$$

$$A^{-1} = \frac{1}{-5} \begin{pmatrix} 5 & -5 \\ -3 & 2 \end{pmatrix}$$

$$= \begin{pmatrix} -1 & 1 \\ 3/5 & -2/5 \end{pmatrix}$$

which gives exactly the same result as we obtained earlier. At the end of the previous section, we suggested that the matrix $\begin{pmatrix} 2 & 1 \\ 4 & 2 \end{pmatrix}$ did not have an inverse. Notice also that the determinant of this matrix is zero. The two facts are indeed connected — if a matrix is to have an inverse then it cannot have a zero determinant.

Matrix Algebra

Let us now consider a simple algebraic relationship using ordinary numbers.

$$ab = c$$

It is easy to see that
$$a = \frac{c}{b}$$
and that
$$b = \frac{c}{a}$$

If the relationship refers to matrices, however, then re-arranging the relationship is not so straightforward. One reason for this is that the re-arrangement above involves division, and so far we have said nothing about division of matrices. In fact, it is not possible to perform division with matrices, so we are going to have to find another method of re-arranging relationships involving matrices.

Suppose A, B and C are matrices such that
$$AB = C$$
and we wish to change this relationship so that just B is on the left hand side (this is called 'making B the subject of the relationship'). Suppose we multiply both sides by A^{-1}, the inverse of A
$$A^{-1}(AB) = A^{-1}C$$
Now as matrix multiplication is associative,
$$(A^{-1}A) B = A^{-1}C$$
and as $A^{-1}A = I$, the identity,
$$IB = A^{-1}C$$
We also know that $IB = B$, so
$$B = A^{-1}C$$
So if we know A and C then we can find the matrix B *provided A has an inverse*.

In a similar fashion, we could make A the subject of the relationship. Multiplying both sides by B^{-1}, the inverse of B
$$(AB)B^{-1} = CB^{-1}$$
Notice carefully the position of the inverse: remember that matrix multiplication is non-commutative
$$A(BB^{-1}) = CB^{-1}$$
$$AI = CB^{-1}$$
$$A = CB^{-1}$$
Again, given C and B, we can find A provided that B has an inverse.

Summarising, if $AB = C$
then $B = A^{-1}C$
and $A = CB^{-1}$

Solutions of Simulataneous Equations

So far, we have spent a lot of time considering matrices without indicating their use. A commonly used application of matrices is in the solution of

simultaneous equations. Suppose we wished to solve the equations
$$2x + 5y = 200$$
$$3x + 5y = 225$$
We can write these equations in matrix form like this:
$$\begin{pmatrix} 2 & 5 \\ 3 & 5 \end{pmatrix} \begin{pmatrix} x \\ y \end{pmatrix} = \begin{pmatrix} 200 \\ 225 \end{pmatrix}$$
For convenience sake, let
$$A = \begin{pmatrix} 2 & 5 \\ 3 & 5 \end{pmatrix}, \quad B = \begin{pmatrix} x \\ y \end{pmatrix}, \quad C = \begin{pmatrix} 200 \\ 225 \end{pmatrix}$$
So the equations take the form
$$AB = C$$
And if we can find the vector B then we will have the value of x and the value of y. In the last section, we saw that if
$$AB = C$$
$$\text{then } B = A^{-1}C$$
Also, we found earlier that if
$$A = \begin{pmatrix} 2 & 5 \\ 3 & 5 \end{pmatrix} \text{ then } A^{-1} = \begin{pmatrix} -1 & 1 \\ 3/5 & -2/5 \end{pmatrix}, \text{ so }$$
$$B = \begin{pmatrix} -1 & 1 \\ 3/5 & -2/5 \end{pmatrix} \begin{pmatrix} 200 \\ 225 \end{pmatrix} = \begin{pmatrix} 25 \\ 30 \end{pmatrix}$$
$$\begin{pmatrix} x \\ y \end{pmatrix} = \begin{pmatrix} 25 \\ 30 \end{pmatrix}$$
or x = 25 and y = 30.
Taking another example, let us solve
$$2x + y = 7$$
$$3x + 2y = 12$$
$$A = \begin{pmatrix} 2 & 1 \\ 3 & 2 \end{pmatrix}, C = \begin{pmatrix} 7 \\ 12 \end{pmatrix}$$
$$|d| = (2 \times 2) - (3 \times 1) = 1$$
$$A^{-1} = \frac{1}{1} \begin{pmatrix} 2 & -1 \\ -3 & 2 \end{pmatrix} = \begin{pmatrix} 2 & -1 \\ -3 & 2 \end{pmatrix}$$
$$\begin{pmatrix} x \\ y \end{pmatrix} = \begin{pmatrix} 2 & -1 \\ -3 & 2 \end{pmatrix} \begin{pmatrix} 7 \\ 12 \end{pmatrix}$$
$$\begin{pmatrix} x \\ y \end{pmatrix} = \begin{pmatrix} 2 \\ 3 \end{pmatrix}$$
The great advantage of using this method is that frequently in the business world we meet situations where the left hand side varies. For example,

suppose an engineering firm produces those invaluable products sproggets and widgets. Each sprogget requires 2 hours casting and 1 hour turning, while each widget requires 1 hour casting and 3 hours turning. Suppose we assume that S sproggets and W widgets are produced: this will require

$2S + W$ hours of casting and

$S + 3W$ hours of turning.

Each week the manager of the department producing sproggets and widgets is informed how much turning and casting time has been allocated to him. Suppose that in a particular week he has been allocated 200 hours of casting and 250 hours of turning — how many sproggets and how many widgets should he produce to use the entire allocation?

$$2s + w = 200$$
$$s + 3w = 250$$

$$\begin{pmatrix} 2 & 1 \\ 1 & 3 \end{pmatrix} \begin{pmatrix} s \\ w \end{pmatrix} = \begin{pmatrix} 200 \\ 250 \end{pmatrix}$$

$$|d| = (2 \times 3) - (1 \times 1) = 5$$

$$A^{-1} = \frac{1}{5} \begin{pmatrix} 3 & -1 \\ -1 & 2 \end{pmatrix} = \begin{pmatrix} 3/5 & -1/5 \\ -1/5 & 2/5 \end{pmatrix}$$

$$\begin{pmatrix} s \\ w \end{pmatrix} = \begin{pmatrix} 3/5 & -1/5 \\ -1/5 & 2/5 \end{pmatrix} \begin{pmatrix} 200 \\ 250 \end{pmatrix} = \begin{pmatrix} 70 \\ 60 \end{pmatrix}$$

So 70 sprockets and 60 widgets would be produced.

Now suppose that the manager is informed how much time has been allocated to him over the next 5 weeks.

Week	1	2	3	4	5
Casting	200	250	225	210	200
Turning	250	200	225	260	200

If we do not use matrices, then we would have to solve five sets of simultaneous equations. However, as we already have the inverse, we can put the availabilities (which would form the right hand side of the equations), into a matrix and solve *all* the equations by matrix multiplication.

$$\begin{pmatrix} 3/5 & -1/5 \\ -1/5 & 2/5 \end{pmatrix} \begin{pmatrix} 200 & 250 & 225 & 210 & 200 \\ 250 & 200 & 225 & 260 & 200 \end{pmatrix}$$

$$= \begin{pmatrix} 70 & 110 & 90 & 74 & 80 \\ 60 & 30 & 45 & 62 & 40 \end{pmatrix}$$

For the next 5 weeks then, his production schedule would be

Week	1	2	3	4	5
Sproggets	70	110	90	74	80
Widgets	60	30	45	62	40

Exercises to Chapter 3

3.1 $A = \begin{pmatrix} 2 & -1 & 0 \\ 3 & 1 & 1 \\ 0 & 4 & -2 \end{pmatrix}$ $B = \begin{pmatrix} 1 & 6 & 2 \\ 4 & 0 & -1 \\ -2 & 3 & 1 \end{pmatrix}$

Find
a) $A + B$
b) $B + A$
c) $A - B$
d) $B - A$
e) AB
f) BA

3.2 If the elements of the matrix A in question 1 are of the form a_{ij}, and the elements of the B matrix are of the form b_{ij}, which elements are denoted by
a) a_{21}
b) b_{32}
c) b_{13}
d) a_{22}

3.3 $P = \begin{pmatrix} 4 & 1 \\ 0 & 2 \\ -1 & 3 \end{pmatrix}$ $Q = \begin{pmatrix} 2 & 1 \\ 3 & 5 \end{pmatrix}$ $R = \begin{pmatrix} 6 \\ 3 \end{pmatrix}$

Determine which of the following products are possible and compute their solutions.
a) PQ
b) PR
c) QP
d) QR
e) RP
f) RQ
g) PQR
h) PRQ
i) RPQ
j) RQP

3.4 Find the inverse of the following matrices.
$A = \begin{pmatrix} 3 & 4 \\ 5 & 1 \end{pmatrix}$ $B = \begin{pmatrix} 4 & 2 \\ 8 & 4 \end{pmatrix}$ $C = \begin{pmatrix} -3 & 5 \\ -1 & -5 \end{pmatrix}$

3.5 Find the inverse of
$$\begin{pmatrix} 3 & 2 & 4 \\ 2 & 1 & 2 \\ 2 & 1 & 4 \end{pmatrix}$$

3.6 Express the following systems of simultaneous equations in matrix form. In each case, find the unknown quantities.

$2x + y = 4$ $3x + y = 13$ $5x + y = 3$
$x + 2y = -1$ $x + 4y = 1$ $6x + 2y = -4$

3.7 Solve the following equations.

(i) $a + b + c = 13$
$2a + b + 2c = 22$
$a + 3b + 4c = 39$

(ii) $2a + b + c + d = 11$
$a + b + 2c + 2d = 17$
$a + 2b + 3c + d = 18$
$2a + 3b + c + 2d = 19$

3.8 To ensure that pigs receive a sufficient vitamin intake, a farmer can mix food additives to pig meal. Three additives are available with the following specifications.

Additive 1. 6 Kg of vitamin A and 4 Kg of vitamin B
Additive 2. 4 Kg of vitamin B, 3 Kg of C and 3 Kg of D
Additive 3. 3 Kg of A, 1 Kg of B. 3 Kg of C and 3 Kg of D.
Put this information into matrix form.

A farmer decides to add 6 containers of additive 1, 8 of additive 2 and 10 of additive 3 to a batch of pigmeal. Put this into an appropriate vector, and deduce the total content of the food.

If a kilo of each vitamin costs £3, £6, £7 and £10 respectively, use an appropriate vector to obtain the cost of each additive, and hence by vector multiplication find the total cost of additives in the batch of feed.

3.9 $X = \begin{pmatrix} 1/6 & 1/3 \\ 1/4 & 1/2 \end{pmatrix}$ and $A = \begin{pmatrix} 10 \\ 45 \end{pmatrix}$

Given that $XT + A = T$, find T.

3.10 On a particular occasion a furniture manufacturer has 115 lengths of hardwood and 120 lengths of softwood. This enables him to make 20 chairs and 25 tables. On another occasion he has 120 lengths of hardwood and 110 lengths of softwood. This enables him to make 30 chairs and 20 tables. If we wish to deduce the hardwood and softwood content of each chair and table state the problem in matrix form and solve it.

3.11 A man has a bottle and a decanter, and the matrix

$R = \begin{pmatrix} a & b \\ c & d \end{pmatrix}$ is such that

a is the quantity of gin in the bottle
c is the quantity of vermouth in the bottle
b is the quantity of gin in the decanter
d is the quantity of vermouth in the decanter.

If the bottle contains 240 ml of gin, and the decanter contains 300 ml of vermouth write down the matrix R.

He now pours ⅓ of the bottle into the decanter. Find the new matrix P that describes the quantities in each container. If Q is the matrix such that

$$QR = P,$$

find the matrix Q.

He tastes the contents of the decanter, and decides the mixture is not strong enough, so he pours half the contents of the decanter back into the bottle. Assuming the quantity tasted is negligible, find the matrix S that describes the quantities in each container.

The mixture in the bottle now suits him perfectly, and he realises that he would like to make more of this mixture in the future. Find the matrix T that describes how to do this, i.e. the matrix T such that

$TR = S$

3.12 Mr. Wealthy and Mr. Rich are investors. Both own stakes in their own company, each others company and in Government securities. Mr. Wealthy always keeps ⅔ of his total wealth invested in his own company (Wealthy Ltd.) and has agreed to keep an amount invested in Mr. Rich's company (Rich Ltd.) equal to ⅓ of Mr. Rich's wealth. Mr. Rich always keeps ¾ of his total wealth invested in his own company, and has agreed to keep an amount invested in Mr. Wealthy's company equal to ⅛ of Mr. Wealthy's total wealth.

Mr. Wealthy has a total wealth of 240 AMU (i.e. arbitrary monetary units), while Mr. Rich has a total wealth of 180 AMU.

Investment in / Investment by	Wealthy Ltd.	Rich Ltd.	Govt. Securities	Total Wealth
Mr. Wealthy				
Mr. Rich				

a) Complete the matrix above.
b) What would the matrix above look like if Mr. Wealthy had 30 AMU and Mr. Rich had 20 AMU invested in government securities?

Chapter Four

Compounding and Discounting

At times, all of us find it convenient to save money. Some very fortunate people cannot help but save as their expenditure is less than their income! For most of us, however, saving involves a conscious effort; we must reduce our consumption if we are to save. As saving involves us in sacrificing consumption today, then we must have a motive. We may save to buy a rather expensive item that we cannot at present afford: or for a deposit to buy a house. We may save as insurance against a rainy day, or to supplement a retirement pension. Now if we deposit our savings in a financial institution such as a building society or bank then, in addition to satisfying our motives, savings yield a bonus to us in the form of the interest earned. In fact, some economists would define interest earned as a reward to us for sacrificing consumption today.

Suppose we deposit a certain sum of money in a bank – then the size of that deposit will grow owing to the interest it earns. It would be useful, then, for us to be able to calculate just what sum of money would be available to us in the future.

Simple and Compound Interest

Do you remember the formula
$$I = \frac{P.T.R.}{100}?$$

There can hardly be an adult alive today who did not meet this formula at school. It tells us how much interest (I) we would earn if we deposited £P in a Bank for T years, and if the bank paid interest rate of R% on deposits. If, then, the bank paid 8% per annum on deposits, then a deposit of £250 left in the bank for 4 years would earn

$$I = \frac{250 \times 4 \times 8}{100} = £80$$

Before we examine the implications of this calculation it will be convenient to modify it somewhat. Rather than expressing the rate of interest as a percentage, let us instead express it as a proportion r. To do this, we divide the percentage rate by 100. So, for example if R = 8% then $r = \frac{8}{100}$ = 0.08. Our formula for calculating interest now becomes

$$I = P.T.r.$$

From now on, we shall use r as the rate of interest. Of course, this would in no way affect the result of our calculation:

$$I = 250 \times 4 \times 0.08 = £80$$

Whichever way we express the rate of interest, we still predict that £250 deposited for four years at 8% per annum would earn us £80. But is this figure correct? We would earn £80 only if we *withdrew* the interest each year. Over the four years we would have earned what is called *Simple Interest*. But suppose we did not withdraw our interest from the bank. If this is so, then the interest on deposit would itself earn interest. We would then be earning what is called *Compound Interest*. Let us now compare the two methods of earning interest on year by year basis.

	Simple Interest			*Compound Interest*			
Year	Deposit	Interest		Deposit	Interest		
1	250	$250 \times 0.08 =$	20	250	250	$\times 0.08 =$	20
2	250	$250 \times 0.08 =$	20	270	270	$\times 0.08 =$	21.60
3	250	$250 \times 0.08 =$	20	291.60	291.6	$\times 0.08 =$	23.33
4	250	$250 \times 0.08 =$	20	314.93	314.93	$\times 0.08 =$	25.19
Total interest earned			80				90.12

So we see that there is a considerable difference between the two methods. Compound interest is the method that is invariably used in the business world, and you would be well advised to forget all about simple interest. If we are going to calculate compound interest on a year by year basis, then the calculation will be tedious to say the least. What we require is a formula for compound interest, and to obtain this we shall consider again our example, though this time from a slightly different angle. We shall calculate the value of the deposit at the end of each year if £P is invested at and interest rate r and left for n years.

Value of deposit at end of year = Value of deposit at beginning of year plus interest earned during the year.

Value of deposit at end of first year =

$$P + rP$$

(deposit at start) (Interest)
$$= P(1+r)$$

Value of deposit at end of 2nd year =
$$P(1+r) + rP(1+r)$$
$$= (1+r)(P+rP)$$
$$= (1+r)P(1+r)$$
$$= P(1+r)^2$$

Value of deposit at end of 3rd year =
$$P(1+r)^2 + rP(1+r)$$
$$= (1+r)^2(P+rP)$$
$$= (1+r)^2 P(1+r)$$
$$= P(1+r)^3$$

So if we call S the value of the deposit (or sum available) after n years.

$$S = P(1+r)^n \qquad \ldots\ldots\ldots\ldots\ldots(1)$$

and the total interest earned would be

$$P(1+r)^n - P \qquad \ldots\ldots\ldots\ldots\ldots(2)$$

Example 1

Now let us repeat our earlier calculation

$P = £250, n = 4, r, = 0.08$

$S = 250 (1.08)^4.$

Some of you will have calculators that can calculate the value $(1.08)^4$. Alternatively, the Compounding Table at the back of this book may be used to give

$$(1.08)^4 = 1.3605$$
$$\text{so } S = £250 \times 1.3605$$
$$= £340.125$$

and the interest earned would be

$$£340.125 - £250 = £90.125$$

which agrees with our earlier, long winded method.

At this point, a few words of warning would be appropriate. Compound interest mounts up very quickly, and yields surprisingly large sums. For example, £100 invested at 8% per annum for 20 years would earn

$$100 (1.08)^{20} - 100$$
$$= 100 (4.6610) - 100$$
$$= £366.10$$

The formula we have derived can also be used to calculate the rate of interest or the number of years, though the calculations are a little more involved.

Example 2

A sum of money is deposited now at 10% per annum. How long will it take for the sum invested to double?

Suppose that the sum invested is P, then after n years we require the sum to be 2P, i.e.

$$P(1+r)^n = 2P$$
$$P(1.1)^n = 2P$$

Dividing both sides by P

$$(1.1)^n = 2$$

We can now search the compounding table to find when $(1.1)^n = 2$. From this table, we see that

$$(1.1)^7 = 1.9487$$
$$\text{and } (1.1)^8 = 2.1436$$

So although the sum invested would not quite have doubled by 7 years it will have more than doubled after 8 years.

Perhaps those of you well used to logarithms may prefer the following method:

$$\text{If } (1.1)^n = 2$$
$$\text{than n. Log}(1.1) = \text{Log } 2$$
$$\text{and } n = \frac{\text{Log } 2}{\text{Log}(1.1)}$$
$$= \frac{0.3010}{0.0414}$$
$$= 7.27 \text{ years}$$

Example 3

£100 is invested now, and we are prepared to leave it on deposit for 15 years. What rate of interest would it be necessary to earn if the sum invested is to grow to £750?

$$100(1+r)^{15} = 750$$
$$(1+r)^{15} = \frac{750}{100} = 7.5$$

Consulting the compounding tables when $n = 15$, we see that

$$(1.14)^{15} = 7.1379$$
$$\text{and } (1.15)^{15} = 8.1371$$

so we must earn a rate of interest somewhere between 14% and 15%. If we use logarithms, we can obtain a more accurate assessment of the rate of interest.

$$100(1+r)^{15} = 750$$
$$(1+r)^{15} = 7.5$$
$$15. \text{ Log}(1+r) = \text{Log}(7.5)$$
$$\text{Log}(1+r) = \frac{\text{Log}(7.5)}{15}$$
$$\text{Log}(1+r) = 0.0583$$
$$(1+r) = 1.1437$$

So the required rate of interest is 14.37%

Now we could use the formula to find the sum we must deposit to achieve a specified sum in the future.

Example 4

We require £10,000 in 15 years time and we can deposit money at 12% per annum. How much must be invest now to achieve this sum?

$$P(1.12)^{15} = 10,000$$
$$P(5.4736) = 10,000$$
$$P = \frac{10,000}{5.4736} = £1,826.95$$

Increasing the sum Invested

So far, we have examined how an initial deposit would grow if it earned compound interest. But suppose we added to the amount deposited at the end of each year. Specifically, suppose we deposited £1000 on the first of January of a certain year, and decided to deposit £100 at the end of each year. If interest is compounded at 10% per annum, then we can use formula (1) to deduce that the

sum on deposit at the end of the first year is
$$1000(1+0.1) + 100$$
sum on deposit at the end of the second year is
$$1000(1+0.1)^2 + 100(1+0.1) + 100$$
sum on deposit at the end of the third year is
$$1000(1+0.1)^3 + 100(1+0.1)^2 + 100(1+0.1) + 100$$
sum on deposit at the end of the nth year is
$$1000(1+0.1)^n + 100(1+0.1)^{n-1} + 100(1+0.1)^{n-2} + \ldots + 100$$

If we generalise the quantities, then we can derive a formula to solve problems like this swiftly and efficiently. If we let P be the initial deposit, r the interest rate and a the amount that we deposit at the end of each year, then after n years the sum available would be

$$S = P(1+r)^n + a(1+r)^{n-1} + a(1+r)^{n-2} + \ldots + a$$

Now it can be shown that this expression is equivalent to[1]

---

1. $S = P(1+r)^n + a(1+r)^{n-1} + a(1+r)^{n-2} + \ldots + a$

Ignore $P(1+r)^n$, then the right hand side forms a geometric progression with a first term $a(1+r)^{n-1}$ and a common ratio $\frac{1}{1+r}$, so using the formula to sum a geometric progression

$$S = P(1+r)^n + \frac{a(1+r)^{n-1}[1-(\frac{1}{1+r})^n]}{1 - \frac{1}{1+r}}$$

$$S = P(1+r)^n + \frac{a(1+r)^{n-1}[1-(\frac{1}{1+r})^n]}{\frac{r}{1+r}}$$

$$S = P(1+r)^n + \frac{a(1+r)^{n-1}[1-(1+r)^{-n}](1+r)}{r}$$

$$S = P(1+r)^n + \frac{a(1+r)^n[1-(1+r)^{-n}]}{r} \qquad S = P(1+r)^n + \frac{a(1+r)^n - a}{r}$$

$$S = P(1+r)^n + \frac{a(1+r)^n - a}{r} \quad \ldots (3)$$

So if we initially deposit £1,000, and add £100 to our deposit at the end of each year, and if interest is compounded at 10% per annum then the sum available after four years is

$$S = 1000(1.1)^4 + \frac{100(1.1)^4 - 100}{0.1}$$

$$= 1000(1.4641) + \frac{100(1.4641) - 100}{0.1}$$

$$= £1928.20$$

We can also use this formula to calculate the sum left on deposit if we withdraw fixed amounts from the bank each year.

Example 5

Suppose we deposit £20,000 at the beginning of a year at 5% per annum compound. We withdraw £2,000 at the end of each year. What would be the sum available after 4 years?

Here we have

$P = 20,000$
$a = -2,000$ (negative because we withdraw)
$r = 0.05$
$n = 4$

$$S = P(1+r)^n + \frac{a(1+r)^n - a}{r}$$

$$S = 20,000(1.05)^4 + \frac{-2,000(1.05)^4 - (-2000)}{0.05}$$

$$S = 20,000(1.2155) + \frac{-2,000(1.2155) + 2000}{0.05}$$

$$S = 24310 - 8620 = £15,690.$$

Formula (3) assumes that a constant amount is added (or withdrawn) from the deposit, and if this amount varies then you must not use the formula. Instead you must use

$$S = P(1+r)^n + a_1(1+r)^{n-1} + a_2(1+r)^{n-2} + \ldots + a_n \quad \ldots (4)$$

where a_1 is the sum added after 1 year, a_2 the sum after 2 years and so on.

Example 6

Suppose we have £20,000 deposited at the beginning of a certain year at 9% per annum compound. At the end of the first year we add £1,000 to the deposit, at the end of the second year we add £2,000 and at the end of the third year we add £3,000 and so on. How much would we have on deposit after 4 years?

$$S = 20{,}000(1.09)^4 + 1000(1.09)^3 + 2000(1.09)^2 + 3000(1.09) + 4{,}000$$
$$= 20{,}000(1.4116) + 1000(1.2950) + 2000(1.1881) + 3000 + (1.09) + 4000$$
$$= 28232 + 1295 + 2376.20 + 3270 + 4{,}000$$
$$= £39{,}173.20$$

If we wish to withdraw varying annual amounts, then we would use formula (4) with negative values for a.

Example 7

If the additions in example 6 became withdrawals, then the amount on deposit after four years would be.

$$S = 20{,}000(1.09)^4 - 1000(1.09)^3 - 2000(1.09)^2 - 3000(1.09) - 4000$$
$$= 20{,}000 - 28{,}232 - 1{,}295 - 2{,}376.20 - 3{,}270 - 4{,}000$$
$$= £17{,}290.80$$

Sinking Funds

We will now suppose that an accountant is instructed to set aside a sum of money at the end of each year to replace an asset.

Using formula 3 i.e.

$$S = P(1+r)^n + \frac{a(1+r)^n - a}{r}$$

this problem involves a zero initial investment. Putting $P = 0$ in the formula

$$S = \frac{a(1+r)^n - a}{r}$$

If we now make a the subject of this formula, we will have an expression telling us how much we must set aside at the end of each year to achieve a specified sum S

$$rS = a[(1+r)^n - 1]$$

$$\boxed{a = \frac{rS}{(1+r)^n - 1}} \qquad \ldots\ldots\ldots\ldots,(5)$$

We have derived what is called the *Sinking Fund* formula.

Example 8

Suppose a machine is expected to last 8 years and its replacement price is estimated at £5000. What annual provision must be made to ensure sufficient funds are available if money can be invested at 8% per annum?

Using formula 5.
$$a = \frac{0.08 \times 5000}{(1.08)^8 - 1}$$
$$= \frac{400}{1.8509 - 1}$$
$$= £470.09$$

So £470.09 deposited at the end of each year would be sufficient to yield the required sum. But suppose (as is more likely) the firm wishes to start the fund now and add to it at annual intervals then formula (5) will not do. We will have to use the expression
$$S = a(1+r)^n + a(1+r)^{n-1} + a(1+r)^{n-2} + \ldots + a(1+r)$$
Now it can be shown that this expression is equal to[2]
$$\boxed{S = a\left[\frac{(1+r)^{n+1} - (1+r)}{r}\right]} \quad \ldots (6)$$
and again rearranging this formula to make a the subject.
$$\boxed{a = \frac{rS}{(1+r)^{n+1} - (1+r)}} \quad \ldots (7)$$

Example 9
Repeat example 8, this time assuming that the sum is invested at the beginning of each year.
Using formula 7
$$a = \frac{0.08 \times 5000}{(1.08)^9 - (1.08)}$$
$$= £435.25$$

Trust Funds and Loan Repayments
Let us suppose that we deposit a certain sum of money now, and from this deposit we wish to withdraw at the end of each year a fixed amount. We will continue to withdraw until nothing is left on deposit. In our formula (3) i.e.
$$S = P(1+r)^n + \frac{a(1+r)^n - a}{r}$$

---

2. Writing this expression backwards
$$S = a(1+r) + a(1+r)^2 + a(1+r)^3 + \ldots + a(1+r)^n$$
$$S = a[(1+r) + (1+r)^2 + (1+r)^3 + \ldots + (1+r)^n]$$
The part in the square bracket forms a geometric progression with a first term $(1+r)$ and common ration $(1+r)$ so
$$S = a\left[\frac{(1+r)(1-(1+r)^n)}{1-(1+r)}\right]$$
$$S = a\left[\frac{(1+r)^{n+1} - (1+r)}{r}\right]$$

The terminal sum S would be zero, and because we are withdrawing a would be negative, so

$$P(1+r)^n - \frac{a(1+r)^n + a}{r} = 0$$

$$P(1+r)^n = \frac{a(1+r)^n + a}{r}$$

$$P = \frac{a[(1+r)^n - 1]}{r(1+r)^n}$$

$$\boxed{P = \frac{a[1 - (1+r)^{-n}]}{r}} \quad \ldots \ldots (8)$$

and

$$\boxed{a = \frac{rP}{1 - (1+r)^{-n}}} \quad \ldots \ldots (9)$$

We shall now examine a few applications of the formulae above.

Example 10

You have decided to set up a trust fund for your Aunt Maud. You require the fund to pay her £2,000 per year for the next 10 years. How much will this fund cost you if money can be invested at 10% per annum compound?

This problem involves finding P, the initial investment, so formula 8 is the one required.

$$P = \frac{2000[1 - (1.1)^{-10}]}{0.1}$$

The Compounding Tables will not help us here, as we require $(1.1)^{-10}$ and not $(1.1)^{10}$. However, the value required can be found in the Discounting Tables printed at the back of this book (You will soon learn why they are called Discounting Tables). From these tables we learn that

$$(1.1)^{-10} = 0.3855, \text{ so}$$
$$P = \frac{2000[1 - 0.3855]}{0.1}$$
$$= £12,290$$

Example 11

Suppose you borrow £3000 at 14% per annum compound, and you wish to repay this loan in 10 annual instalments. How much must you repay each year?

If you think carefully about this problem, you will realise that it is equivalent to asking how much can be withdrawn at the end of each year if

£3000 is invested now at 14% per annum. So formula 9 is appropriate to this problem.

$$a = \frac{0.14 \times 3000}{1 - (1.14)^{-10}}$$

Using the discounting tables, $(1.14)^{-10} = 0.2697$

$$a = \frac{0.14 \times 3000}{1 - (0.2697)}$$
$$= £575.10$$

The Concept of Present Value

Suppose you were offered the choice of receiving £1000 now or £1000 in twelve month's time — which would you choose? It is almost certain that you would take the money now, even if you had a cast iron guarantee of receiving the money in the future. It would appear that we have a strong preference for holding cash now as against receiving cash in the future, and economists call this preference 'Liquidity Preference'. Now why is this preference so universally held? Almost certainly, inflation will have something to do with it. After all, if prices are rising then £1000 in 1 year's time will buy less than it will buy now, and so it will have less value than it has now. During inflation, then, it would make sense to take the £1000 now! But suppose (wishfull thinking) we had stable prices — we would still almost certainly choose to take the money now. Why?

The great advantage of taking the money now is that it can be invested, earn interest, and grow. Our £1000 invested now at 10% would grow to

£1000 (1.1) = £1100 in one years time.

and to £1000 $(1.1)^2$ = £1210 in two years time.

Given stable prices, then, we should be indifferent between £1000 now, £1100 in one years time and £1210 in two years time. In other words, £1000 receivable today has the same value as £1100 receivable in one year's time and £1210 receivable in 2 year's time. So we can now see why we would prefer the £1000 now: if £1000 now is worth £1100 in one years time, it follows that £1000 in one years time has a *present value* of less than £1000.

We have introduced a very important concept — that of present value, and this concept needs defining carefully. The present value of a sum of money receivable in the future is the sum you would be prepared to accept now, rather than have to wait for it. We use the interest earning capacity of money to enable us to calculate the present value. As £1000 invested at 10% per annum would grow to £1100 in one year's time we would say that £1100 in one years time has a present value of £1000. We reduce *or discount* the value of a sum receivable in the future to find its present value, and the *discount factor* that we use to do this is the current rate of interest. How can we do this? The first formula we obtained in this chapter was used to calculate how a sum invested now would grow under compound interest.

$$S = P(1+r)^n$$

If S is the sum receivable in the future, then P must be its present value.

$$P = \frac{S}{(1+r)^n}$$

or $\boxed{P = S(1+r)^{-n}}$ 10

The quantity $(1+r)^{-n}$ is the *discounting factor* reducing the value of the sum. (hence the name discounting tables for values of $(1+r)^{-n}$)

Example 12

What is the present value of £1000 receiveable in 2 year's time if money can be invested at 10% per annum compound?

$S = 1000$
$r = 0.1$
$n = 2$
$P = 1000 (1.1)^{-2}$

Using the Discounting Tables

$P = 1000 \times 0.8264$
$= £826.40$

So £1000 receivable in 2 year's time has a present value of £826.40 because if we invest £826.40 now at the current interest rate it would grow to £1000 in 2 years. We should be indifferent between receiving £1000 in 2 year's time and £826.40 now.

The Present Value of a Stream of Earnings

Let us suppose that we have been promised £1000 in one year, £2000 in two years, £4000 in three years and £3000 in four year's time. We would call this a stream of earnings, and if we wished to find the present value of such a stream we would find the sum of the individual present values. If the current rate of interest is 12%, then the present value would be

$$1000 (1.12)^{-1} + 2000 (1.12)^{-2} + 4000 (1.12)^{-3} + 3000 (1.12)^{-4}$$

It is conventional to perform this calculation in a tabular form

Year	Earning	Discount Factor $(1.12)^{-n}$	Present Value
1	1000	0.8930	893.00
2	2000	0.7972	1594.40
3	4000	0.7118	2847.20
4	3000	0.6355	1906.50
			7241.10

So the stream of earnings has a present value of £7241.10. In other words, if we deposited £7241.10 now at 12% per annum then we could draw £1000 at the end of the first year, £2000 at the end of the second year, £4000 at the end of the third year and £3000 at the end of the fourth year. The value of the deposit would then be zero (you should use formula 4 to prove this yourself). So if you were offered an asset which yielded the above stream of earnings and the current interest rate was 12% per annum, then you should be prepared to pay £7241.10 for it.

Now it is perfectly possible to have a constant stream of earnings, and an asset that yields a constant stream is called an *annuity*. We could use the same method as above to calculate the present value of an annuity, but it would be easier to use formula 8, which would give exactly the same results.

Example 13.
You are offered an annuity that would yield £1500 a year for 10 years. How much would you be prepared to pay for it if interest is compounded at 8% per annum?

$$P = \frac{a[1 - (1+r)^{-n}]}{r}$$

$$= \frac{1500 [1 - (1.08)^{-10}]}{0.08}$$

$$= £10,065$$

Example 14
Repeat example 13, but this time assume the rate of interest is 12% per annum.

$$P = \frac{1500 [1 - (1.12)^{-10}]}{0.12}$$

$$= £8,475$$

The last two examples prove an important point: if the rate of interest rises then the value of a future stream of earnings falls.

Suppose we now consider an annuity that pays a stream of earnings indefinitely – such an asset is called a *perpetual annuity*. How would we find the present value of such an asset? In the formula

$$P = \frac{a[1 - (1+r)^{-n}]}{r}$$

the value of n will be infinity. So $(1+r)^{-n}$ would be zero, and the formula reduces to

$$P = \frac{a}{r}$$

Example 15
An annuity yields £1200 per year. What is its present value if the current rate of interest is 20% per annum?

$$P = \frac{1200}{0.2} = £6,000$$

It is easy to see that the present value must be £6000, for if £6000 was invested at 20% per annum then the *simple* interest earned would be £1200 per year. So we could withdraw £1200 per year and leave the capital intact.

Finally, we can use the concept of present value to estimate the market price of fixed interest stock.

Example 16

Ruritarian Transport Stock has a nominal value of £100, and pays a dividend of 7%. Estimate the market price if the stock has eight years to run to maturity, and the current market rate of interest is 10%.

Holding this stock would yield £7 per year for eight years. This has a present value of

$$P = \frac{7[1 - (1.1)^{-8}]}{0.1} = £37.34$$

Also, the stockholder would receive £100 in eight year's time when the stock matures. This has a present value of

$$100(1.1)^{-8} = £46.65$$

So the estimated market price is £37.34 + £46.65 = £83.99, because if this sum was invested at current market rates of interest, then cash flow could be obtained identical to that from the stock.

Exercises to Chapter 4

4.1 If £3250 is invested now at 11% per annum compound, what sum would be available in 7 year's time?

4.2 Complete the table below

	Years required for investment to		
	Double	Treble	Quadruple
Compound Rate 8%			
12%			
16%			
20%			

4.3 What compound rate would cause £3,265 to grow to £5,776 in six year's time?

4.4 How much must be invested now if you require £12,500 in 5 year's time and if money can be invested at 10% per annum?

4.5 Suppose £9,500 is invested on the first of January of a certain year at 12% compound and £800 is withdrawn at the end of each year. How much would remain after 12 years?

4.6 Suppose £7500 is invested on the 1st of January at 9% per annum compound. Withdrawals are: £1000 at the end of the first year, £1200 at the end of the second year and £2000 at the end of the third year. How much would then remain?

4.7 A machine costing £12,500 now will need replacing in 6 year's time.
  a) Estimate its replacement price if the rate of inflation is 11% per annum.
  b) How much must be set aside
    (i) at the end of each year
    (ii) at the beginning of each year
   to replace the machine?

4.8 If money can be invested at 9% per annum, how much must be invested now to yield an income of £5000 per year, paid at the end of each year for eight years?

4.9 Suppose £6500 is borrowed at 18% per annum compound. Find the annual repayment necessary to pay off the loan in 12 years.

4.10 Given a discount rate of 14% per annum, find the present value of an annuity which yields
  a) £650 per year for 8 years
  b) £650 per year for ever.

4.11 Estimate the market price of a holding of 7% Treasury Stock with a nominal value of £125,000 and 4 years to run to maturity, if the current rate of interest is 10%.

4.12 A machine costs £150,000 and its estimated running costs over its life of 5 years are

year	Running Cost
1	1250
2	2250
3	3000
4	3100
5	3200

Assume all running costs are paid at the end of each year, and that money can be invested at 12% per annum compound.
  a) How much must be set aside to cover running costs?
  b) How much must be set aside to cover running costs and replace the machine?
  c) How much must be set aside to cover running costs and replacement idenfinitely?

Chapter Five

Investment Appraisal

There is little doubt that one of the outstanding features of the last hundred years is the development of new technologies and the application of those technologies to industrial and commercial uses. If you consider such an everyday matter as the passage of information, it could be effected in 1800 only by letter or by personal contact. Today we have a full range of means of communication available to everyone — telephone, telegram, radio, television, telex and many others, most of which did not exist until after 1900. So true is this that many of our major firms now have Communications Sections responsible for developing channels of communication and for choosing those technologies which best meet the communications needs of the firm.

As you can guess, this technological revolution has presented the industrialist with a new set of problems. Not only does he have to decide what to produce, he now has to decide also which of many alternative methods he is going to use. You may, of course, argue that this is not really a new situation. Choice and alternatives have always been with us. The new dimension lies in the number of alternatives and in the cost involved. If you are considering expenditure of only a few hundred pounds, intuition or rule of thumb methods of selection may do little harm. But if expenditure runs into millions of pounds such methods may be catastrophic and lead to bankruptcy.

This, then, is what *investment evaluation* is all about. We have to devise criteria for choice and methods of looking at investment which will enable the business man to decide which of many different investments is best for him — or indeed, if any one of them should be undertaken.

But what criteria should we use? It seems self evident that, since a large expenditure is involved, a financial return will be required, and that the magnitude of that return should be our criterion. Yet, as we saw in the last chapter, we have to be careful. Firstly because two projects yielding different returns may also have involved different capital expenditures and thus are not directly comparable. We cannot offhand say that a project costing £70,000 and yielding a return of £4550 each year is better than one costing £55,000 and yielding an annual return of £3850. In fact, neither is as profitable as putting the money into a Building Society account at 9% interest. It is the rate of return rather than the absolute amount which is important. Secondly, as you know, the timing of the returns is also important. An investment yielding £10,000 a year for three years followed by £2000 a year for three years is very different from one yielding £2000 a year for the first three years and then £10,000 a year for three years.

So it would seem that in developing criteria for investment evaluation, at least three things have to be borne in mind:

a) the financial return
b) the rate of return
c) the timing of the return.

Methods of Investment Appraisal

We will first look at methods of assessing the relative profitability of investment projects which evaluate the returns from the investment but ignore the timing of the returns. There are certain assumptions implicit in using one of these methods:

i) that money received at some future date is worth as much as money received now.

ii) that we can assess with some accuracy both the costs and the net revenues of the different projects.

Furthermore, to keep the analysis fairly simple we will assume that we will not gain by lodging our money on deposit with a bank or a Building Society. Remember, though, that any business-man can do this and assure himself of an income. This income is, then, a minimum below which the cash return from an investment project must not fall.

Given these assumptions, a businessman may decide to adopt as his criterion one of two things:

either i) how long it will take him to recover his initial outlay, – the payback period.

or ii) the relationship between the profits made and his initial outlay – the rate of return.

Payback Period Method

The *payback period* is defined as the number of years it takes the cash proceeds from the investment to equal the initial outlay:

Example:

Omega Engineering is considering its investment programme. It has to assess four different projects each having the same initial capital cost of £10,000. The Financial Director informs the Board that only one of the projects can be financed, and provides the following estimates of potential income. He has taken account of all expenses such as wages and the cost of raw material but has made no provision for depreciation.

		Net Cash Flow		
Project	Cost	Year 1	Year 2	Year 3
	£	£	£	£
A	10000	10000	–	–
B	10000	5000	5000	5000
C	10000	1500	4500	6000
D	10000	6000	8000	5000

Project A is a short term investment. It has a life of one year only, but during that year the net cash flow is just sufficient to cover the initial outlay. Its payback period is, therefore, one year.

Project B returns a steady cash flow of £5000 a year during its life. Two years is sufficient to recover the initial outlay of £10,000 and its payback period is therefore two years.

Project C takes time to build up its potential. During the first two years it earns £6000 in total, £4000 short of the initial cost. If we assume that the £6000 earned in year 3 is earned at an even rate throughout the year, (£500 a month), it takes a further eight months to accumulate this £4000. The payback period is two years eight months, i.e. 2.67 years.

Project D is a high yielding project in its first two years. Again assuming an even flow of earnings the payback period is 1.5 years.

We can now rank the four projects according to the criterion we have adopted – the length of the payback period.

Project	Payback Period (years)	Rank
A	1	1
B	2.0	3
C	2.67	4
D	1.5	2

Project A would be the investment most favoured. Yet, if you think about it, this seems to be an illogical choice. Firstly, we ignore any income received after the payback period. The total income from every other project is greater than that from Project A, yet they are rejected. Thus the method is biassed in favour of those projects with a high cash return in their early years. In times of stringency, when cash flow is important, this is understandable, but very many investments which will ultimately produce a high and steady cash flow for many years take time to yield results. The stability such investments add to a firm's position merits more consideration than this method gives it.

Secondly, we ignore the timing of the cash flow from the point of view of the cost of funds. If the rate of interest is 10%, £10,000 received after one year from project A has a present value of only £9,090.91. Yet we have to invest £10,000 to receive it. It is just not worth while! Put another way, if we lodged the £10,000 in a bank at 10% after one year we would have £10,000 $(1 + 0.1)$ = £11,000. Investing in project A leaves us with only £10,000 (assuming the machine is worth nothing). Consider now project D. The present value of the cash flows over the three years is

6000(.0991) + 8000(.8264) + 5000(.7513) = 5454.60 + 6611.20 + 3756.50
= £15822.30

and for a cash flow of this present value we sacrifice only £10,000. Serious doubts must be expressed as to whether payback period calculations will lead us to the correct investment decision.

In spite of this, research work carried out by Merritt and Sykes in 1964 showed that 78% of firms here and abroad were using this method. Ten years later an English Institute Research Committee found that it was still the most popular single criterion for investment decisions. Why should this be so?

Well, it is, of course, very easy to understand and apply, and the data it uses is readily available to managers. It seems, too, that in emphasising high early cash flows it satisfies the desire of management for high liquidity. We must admit, and this is something we all tend to forget, that the further we project our potential cash flows into the future, the greater is the risk that the returns we anticipate will not be realised. The manager usually wants to be able to point to a record of success, so he will tend to favour shorter term, low-risk investments in which the initial outlay is recouped quickly.

We cannot help feeling that the continued use of this technique indicates some of the pressures influencing management decisions in the twentieth century, but at the same time it indicates that managers as a body sometimes act in an apparently illogical manner. In view of this might it not be better to look at the total potential income from an investment. We could, then, consider the average annual return as a percentage of cost, either before or after depreciation is allowed for.

Average Annual Rate of Return Techniques

We will firstly consider the concept of gross returns — that is, returns without considering the cost of depreciation. The *gross average annual rate of return* is defined as:

> The average proceeds per year over the whole life of the capital expressed as a percentage of the initial capital cost.

Using the same data that we used to calculate payback period we get the following results:

Project	Capital Cost	Total Cash Flow	Life (years)	Average Cash Flow	% of Capital Cost	Rank
A	10,000	10,000	1	10,000	100	1
B	10,000	15,000	3	5,000	50	3
C	10,000	12,000	3	4,000	40	4
D	10,000	19,000	3	6,333	63.3	2

You will notice that this method has produced exactly the same ranking as the payback period, but this is purely coincidential. If we were to increase the revenue for, say, project B to £11,000 in the third year, it would not affect the payback period, but it would raise the total cash flow to £21,000 and annual average cash flow to £7,000 which is a return of 70% of capital cost. Thus it would be ranked 2.

This method is an improvement. It is still easy to understand, but now it does take account of revenue over the whole life of the asset. Moreover, in using this concept of return on capital employed it is using a yardstick which is familiar to most businessmen and accountants. But it is still a criterion which ignores the critical factor of the timing of the cash flows. It is possible to imagine a firm being forced into liquidation before the very high cash flows anticipated in the future can materialise. We have no indication also of the time span of the returns. A return of 20% for one year only would be equated with a return of 20% each year for five years.

Once again the concept is open to criticism. Yet it is often found in practice – usually as a means of checking the results of more sophisticated techniques. Would we perhaps be better if we were to consider profit rather than cash flow? We will use profit in the normal sense of the net gain to the company after maintaining the value of its assets intact. We will also assume that no investment has any cash value at all once its life has passed. This means that we have to reduce the total cash flow by the cost of capital (£10,000) in order to find the total net cash flow. We will also assume that the depreciation is charged by equal annual instalments.

Project	Cost	Total Cash Flow	Depreciation	Net Cash Flow	Net Average Cash Flow	Annual % of Capital	Rank
A	10,000	10,000	10,000	0	0	0	4
B	10,000	15,000	10,000	5,000	1667	16.67	2
C	10,000	12,000	10,000	2,000	667	6.67	3
D	10,000	19,000	10,000	9,000	3000	30.00	1

Now this is much better. At last we have a ranking indicative of benefit to the firm. Project A which only just recovers its cost is no longer first choice. In fact, it is last choice, and is shown as not being profitable. But are we being fair? Because the project lasts for one year only we are charging the whole cost of capital against that years income. The longer the life of capital the lower is the average annual depreciation charge and the more likely it is that the average annual net cash flow will be positive. Many people argue that the very basis of our calculation is wrong. If we charge depreciation against the cash flow we ought not to calculate that cash flow as a percentage of initial capital cost. Since reinvestment of the depreciation provision makes the capital self-liquidating we should express the average annual net cash flow as a percentage of the average capital employed. In this case that would be as a percentage of £5,000.

Whatever may be the merits of these arguments, all the methods we have looked at so far suffer from one glaring defect – they ignore the timing of the cash flows. They assume that, say, £5,000 received in five year's time has the same value to the industrialist as £5,000 received next year. As we have seen, this is most certainly not so. It is to this concept that we must now turn our attention.

Discounted Cash Flow Techniques

As we saw in the last chapter money received now is preferred to money received in the future, and interest is paid to overcome this time preference. We developed a method of calculating the present value of a sum of money receivable in the future. If you think back you will remember that the present value of such future receipts depends on two factors:

a) the time period involved. The longer the interval before receipt of the cash, the lower is the present value.
b) the rate of interest. The higher the rate of interest, the lower will be the present value.

We developed the formula for calculating present value – the present value of £a receivable in n years is:

$$\text{Present Value} = a(1+r)^{-n}$$

where r = the rate of interest

So that you will not have the cumbersome task of calculating $(1+r)^{-n}$ the discount tables at the back of the book give you the present value of £1 for values of n up to 20 years and values of r up to .35.

Now if you think about it this problem of present value is precisely the problem the businessman is facing. In investing he is spending money *now* in order to receive a flow of cash at various times *in the future*. This too is the weakness of all the methods in the previous sections. It is futile to argue that if an investment has returned £5,000 at the end of year one and a further £5,000 at the end of year two the businessman has recouped the capital cost of £10,000. He has, in fact, recouped only the present value of the two cash payments, namely, given a rate of interest of 14%

$$5000(.8772) + 5000(.7695) = £8233.50$$

In considering an investment we may find ourselves in two completely different positions. We may have ample funds for investment and merely wish to know if a particular project will yield a rate of return equal to the cost of providing funds. In such cases we will use a technique usually known as the *Net Present Value Method*. Often, however, funds are limited, and we are faced by a number of projects competing for limited funds. We wish to invest in that project yielding the highest rate of return. To solve this problem we will use a technique referred to as the *Internal Rate of Return Method*.

Net Present Value Method

You will remember that in the last chapter you learned how to calculate the present value of a sum of money receivable at some date in the future. You will understand, too, that no manufacturer is going to use his money to buy capital equipment unless there is every prospect that the return on that capital will be at least equal to the return he could get by putting his money into a building society or bank account. Let us bring these two concepts together to see how they enable a businessman to make an investment decision.

Example:

Mr. Arnold White, a market gardener is considering whether to buy a new cultivator to help him in his business. The cultivator would cost him £6,000 but would enable him to undertake contracts which he estimates will be worth £3,000 a year after all expenses have been paid. Unfortunately the cultivator has a life of only three years and will have no scrap value at the end of that time. The current rate of interest is 12%. Should he undertake the investment?

Here Mr. White has several problems to consider. It is not just a question of the income he will receive. Obviously £3,000 a year is a much higher income than he would receive if he put his money on deposit at 12% – or even 20%. But against this is the fact that if he places his cash on deposit he will still have his £6,000 at the end of three years, whereas if he buys the cultivator, he will be left with only a worthless machine. If the investment is to be worthwhile then, he must consider not only the annual income, but also whether he can recoup the capital cost of the machine. The best way of assessing this is to calculate the present value of the annual cash flows.

Now, in doing this, do not calculate $(1+r)^{-n}$ over and over again. At the end of the book you will find discount tables giving the value of $(1+r)^{-n}$ for all values of r from .01 to 0.35 and for all values of n from 1 to 20. Do make sure you understand how to use these tables.

We can now calculate the present value of the cash flows.

End of Year	Cash Flow	Discount Factor $(1+.12)^{-n}$	Present Value
1	3000	.8929	2678.70
2	3000	.7972	2391.60
3	3000	.7118	2135.40
			£7205.70

Thus the cash flows have a total present value of £7205.70. How can you interpret this? Well, on buying the machine we are in fact buying an annuity which will yield an income of £3000 a year for three years. You could use your annuity formula to calculate the cost of such an annuity if the rate of interest were 12%. But this is what we have calculated when we have calculated the present value of the cash flows. So we obtain an annuity worth £7205.70 by spending £6000 on the cultivator. This is quite a bargain by any standards.

The investment has a net present value of £7205.70 less the cost of the cultivator.

$$\text{Net Present Value} = £7205.70 - £6000 = £1205.70$$

Now if you think about it, it is obvious that the net present value of an investment will depend on what you think the net profit will be year by year. If costs are rising and hence profits are falling, net present value is reduced. Profit figures for the future are in fact only estimates and can be materially affected by the optimisms or pessimism of the businessman.

Example:

Shortly before he buys the cultivator Arnold White is shocked to discover that his annual wage bill has risen by £500 due to a wage award. Being a pessimist he believes that the wage bill will continue to rise year by year, and that he will be able to cover only a part of these increased costs by price increases. He revises his ideas and comes up with the following forecast of earnings:

Year 1   £2500   Year 2   £2400   Year 3   £2350

He wonders if he should really buy the cultivator in these circumstances.

Year	Earnings	$(1+.12)^{-n}$	Present Value
1	2500	.8929	2232.25
2	2400	.7972	1913.28
3	2350	.7118	1672.73
		Present Value =	5818.26
		Less	6000.00
		Net Present Value =	−£ 181.74

The Net Present Value is negative. White is being asked to pay £6000 for an income which to him is worth only £5818.26 today. He would do better to hold on to his money and deposit it in a bank.

Net Present Value can also be affected by changes in the rate of interest. This should be obvious if you remember that the value of our discount factor $(1+r)^{-n}$ depends on the value of r. Let us see how changes in r will affect the investment decision.

Example:

Just as he is getting desperate Arnold White is told that the rate of interest is about to fall from 12% to 9%. Does this affect his decision not to buy the cultivator?

His calculations now appear as follows:

Year	Earnings	$(1+.09)^{-n}$	Present Value
1	2500	.9174	2293.50
2	2400	.8417	2020.08
3	2350	.7722	1814.67
			6128.25
		Less	6000.00
		Net Present Value	£ 128.25

As you can see, the fall in the rate of interest has raised the present value of the cash flows and the investment again becomes worth while. It is logical that this should be so. With the fall in the rate of interest, the alternative use of money (putting it in a bank or building society) yields a lower return. The investment is, therefore, relatively more profitable.

Although Arnold White is only a small market gardener he has taught us a great deal. Summarising what we have learned so far:

a) Net Present Value = Present Value of the cash flows less the cost of the investment.

b) Net Present Value may be positive or negative. If it is positive the investment is worthwhile; if it is negative the investment should not be undertaken.

c) A fall in interest rates or a rise in net earnings will increase net present value. A rise in interest rates or a fall in net earnings will reduce net present value.

As you will probably have appreciated already, this method can be used to compare the relative merits of several investments.

Example:

A salesman calls in to see Arnold White. He hopes to be able to sell to him a bigger and better cultivator. There is one on the market costing £8000, which, because it does the job more quickly should ensure that returns are higher. Arnold does a quick estimate and comes up with the following figures:

	Machine A	Machine B
Cost	6000	8000
Earnings:		
Year 1	4000	4800
2	4000	6400
3	4000	4000

He wonders which cultivator he should buy. It is no use his buying both. The current rate of interest is 10%.

		Project A		Project B	
Year	$(1+.1)^{-n}$	Cash Flow	Present Value	Cash Flow	Present Value
1	.9091	4000	3636.40	4800	4363.68
2	.8264	4000	3305.60	6400	5288.96
3	.7513	4000	3005.20	4000	3005.20
			9947.20		12657.84
	Less Cost of Capital		6000.00		8000.00
	Net Present Value		£3947.20		£4657.84

We now face a tricky problem. Both machines are profitable but we have not considered any criterion to help us decide which is the better. Many people would say that all they are interested in is contribution to profit. If this is so, machine B certainly adds more to profit than does machine A. We could make the choice by a direct ranking of the net present value. But is this really correct? Perhaps we should be interested in the efficiency of the capital we employ and look at the percentage return. Direct comparison will give us this only if the cost of capital were the same in both cases. Since it is not, we must turn the net present value into a profitability ratio index by expressing it as a proportion of the cost of capital. Thus:

	Machine A	Machine B
Capital Cost	6000	8000
Net Present Value	3947.20	4657.84
Index	$\dfrac{3947.20}{6000.00} = .658$	$\dfrac{4657.84}{8000.00} = .582$

Now the boot is on the other foot. In terms of present value, each pound spent on machine A contributes 65.8 pence to profits, whereas each pound spent on machine B contributes only 58.2 pence.

We will not try to tell you which of these two criteria is the better. Each firm will have its own policy and will have laid down its own criteria. Provided they have done so for logical reasons and are consistent in their policies, there is little more to be said.

Internal Rate of Return Method

If you think back over the methods of investment evaluation we have considered, you will quickly see that one thing is missing. Nowhere have we actually calculated the *rate* of return. Net present value measures the contribution to profits: the profitability index measures the contribution of each pound of capital invested; even the so-called rate of return measures only the level of profits as a percentage of capital cost. As yet we have not measured the rate of return in terms of interest.

This is precisely what the internal rate of return does. Many firms find the method suitable because they are not only interested in an investment being profitable – they want to know how profitable. Thus the internal rate of return is an extremely useful way of comparing several investments when we can afford to finance only one of them. There is more than this however. In all our examples we have been given a rate of interest to enable us to calculate the discount factor. Now there are times when an appropriate rate of discount cannot be determined. When this is so, the only way we can judge whether an investment is worthwhile is by assessing its rate of profit as compared with some predetermined standard.

In principle the technique is very easy to understand. We attempt to determine the rate of discount which will equate the present value of future cash flows with the capital cost of the investment. In practice it is not so easy. To find the true discount rate we have to proceed by trial and error until we come across it. Let us see how it works:

Example:
Arnold White has bought his cultivator for £6000 and believes that his

returns for the next three years will be £2500, £3000 and £2000. Evaluate this investment by calculating the internal rate of return.

One way of tackling the problem would be to calculate the present value of the three cash flows using every rate of discount from 1% upwards, until we find one that equates present value with capital cost. It is not a method we recommend! Surely it would be better to try to estimate the rate of return and start our search with that estimated figure. So we proceed as follows. The total cash flow over the life of the project is £2500 + £3000 + £2000 = £7500. Since the cultivator cost £6000 this leaves us with a surplus of £1500, i.e. $\frac{1500}{6000} \times 100 = 25\%$ of capital cost. Quite simply this could be looked on as 8.33% a year. It would be better to start our search here rather than at 1% or by pure guesswork. We know, however, that calculating the rate of return in this way *underestimates* the internal rate of return so we might as well allow for this. A good working rule is to increase this estimate by about a third to a half. So we could estimate the internal rate of return to be somewhere in the region of 8.33 × 1.33 to 8.33 × 1.5 or between 11.1 and 12.5. So we will start our search for the true internal rate of return using a rate of discount of 11%.

Year	Cash Flow	$(1+.11)^{-n}$	P.V.	$(1-.12)^{-n}$	P.V.	$(1+.13)^{-n}$	P.V.
		$r = .11$		$r = .12$		$r = .13$	
1	2500	.9009	2252.25	.8929	2232.25	.8850	2212.50
2	3000	.8116	2434.80	.7972	2391.60	.7831	2349.30
3	2000	.7312	1462.40	.7118	1423.60	.6931	1386.20
Present value			6149.45		6047.45		5948.00

Trying $r = .11$ first we find the present value to be £6149.45, too high a figure. So the rate of discount we have chosen is too low. When we increase r to .12 the present value falls to £6047.45, still too high but we are getting very close. Moving to $r = .13$ we find the present value is £5948 − too low. So the true internal rate of return lies between 12% and 13%. For most purposes this would suffice but we can find the exact figure if we wish by interpolation. The technique is simple:

$$\text{Internal rate of return} = 12\% + \frac{(\text{P.V. when } r = .12) - (\text{capital cost})}{(\text{P.V. when } r = .12) - (\text{P.V. when } r = .13)}$$

$$\text{I.R.R.} = 12\% + \frac{6047.45 - 6000}{6047.45 - 5948.00}$$

$$= 12 + \frac{47.45}{99.45} = 12.477\%$$

We can easily prove that this is the true internal rate of return by calculating the discount factors for $r = .12477$. (These, of course are not in your printed tables, but can easily be obtained on your calculator).

Year	Cash Flow	$(1.12477)^{-n}$	Present Value
1	2500	.88907	2222.675
2	3000	.79045	2371.35
3	2000	.70276	1405.52
			£5999.55

There is a 45 pence error due to approximating the discount factor to five decimal places and because of our approximating the rate of return to three decimal places.

On the surface 12.477% may seem a high rate of return. After all, you will say, profits are only £1500 and if we invested £6000 at only 9% we would earn £1620 at simple interest and £1770 at compound interest in three years. What is wrong then. Well, if you argue like this, what is wrong is your understanding of what we are doing. Look at it like this.

If we invest £6000 at 12.477%, at the end of the first year we have £6000 × 1.12477 = £6748.62. We now withdraw £2500 leaving £4248.62. By the end of the second year interest has increased this to £4248.62 × 1.12477 = £4778.72. We now withdraw £3000 leaving £1778.72. At the end of the next year this will amount to £2000.65 when we withdraw £2000 closing the account. Again we get a slight error due to rounding. It is this type of situation we are considering when we calculate the internal rate of return. We are in fact calculating the rate of interest which will make the present value of an annuity yielding unequal cash flows equal to a predetermined sum.

Let us conclude this chapter by examining the results produced by the various methods we have discussed.

Example:

The board of M.A.T. Ltd are considering three alternative investment policies presented to them by their planning team.

Project A costs £25000 and has a life of ten years. The estimated cash flows are £5000 in year 1, rising by £500 a year to £7000. Returns then remain steady for the life of the asset.

Project B costs £8000 but has a life of only 4 years. It yields £4000 during each of the first two years, £2000 in year 3 and £1000 in year 4. It is estimated, however, that it will have a scrap value of £1000.

Project C costs £10000, also lasts for 4 years and it is believed that it will yield £5000 in year 1, £4000 in year 2, and £3000 in each of the remaining two years.

The Board has no set policy but wishes each project to be evaluated and ranked in order of preference using

a) Payback period   c) Profitability Index
b) Net Present Value (N.P.V.)   d) Internal rate of return (I.R.R.)

The rate of interest is 10%.

Payback Period

Project	Cost	Payback Period	Rank
A	25000	4.29	3
B	8000	2.00	1
C	10000	2.33	2

Net Present Value

Project A

Year	Cash Flow	$(1+.1)^{-n}$	Present Value
1	5000	.9091	4545.50
2	5500	.8264	4545.20
3	6000	.7513	4507.80
4	6500	.6830	4439.50
5	7000	.6209	4346.30
6	7000	.5645	3951.50
7	7000	.5312	3592.40
8	7000	.4665	3265.50
9	7000	.4241	2968.70
10	7000	.3855	2698.50
			38860.90
		Less	25000.00
		Net Present Value	£13860.90

Project B

Year	Cash Flow	$(1+.1)^{-n}$	Present Value
1	4000	.9091	3636.40
+	4000	.8264	3305.60
3	2000	.7513	1502.60
4	2000	.6830	1366.00
			9810.60
		Less	8000.00
		Net Present Value	£1810.60

Project C

Year	Cash Flow	$(1+.1)^{-n}$	Present Value
1	5000	.9091	4545.50
2	4000	.8264	3305.60
3	3000	.7513	2253.90
4	3000	.6830	2049.00
			12154.00
		Less	10000.00
		Net Present Value	£2154.00

Summary of Net Present Value

Project	N.P.V. (£)	Rank
A	13860.90	1
B	1810.60	3
C	2154.00	2

Profitability Index

Project	Cost	N.P.V.	Index = $\frac{\text{N.P.V.}}{\text{Cost}}$	Rank
A	25000	13860.90	.554	1
B	8000	1810.60	.226	2
C	10000	2154.00	.215	3

Internal Rate of Return

Project A

Year	Cash Flow	$(1 + .2)^{-n}$	P.V. (r = .20)	$(1 + .21)^{-n}$	P.V. (r = .21)
1	5000	.8333	4166.50	.8264	4132.00
2	5500	.6944	3819.20	.6830	3756.50
3	6000	.5787	3472.20	.5645	3387.00
4	6500	.4823	3134.95	.4665	3032.25
5	7000	.4019	2813.30	.3855	2698.50
6	7000	.3349	2344.30	.3186	2230.20
7	7000	.2791	1953.70	.2633	1843.10
8	7000	.2326	1628.20	.2176	1523.20
9	7000	.1938	1356.60	.1799	1259.30
10	7000	.1615	1130.50	.1486	1040.20
			25819.45		24902.25

$$\text{I.R.R.} = 20 + \frac{819.45}{25819.45 - 24902.25} = 20.89\%$$

Project B

Year	Cash Flow	$(1 + .21)^{-n}$	P.V. (r = .21)	$(1 + .22)^{-n}$	P.V. (r = .22)
1	4000	.8264	3305.60	.8197	3278.80
2	4000	.6830	2732.00	.6719	2687.60
3	2000	.5645	1129.00	.5507	1101.40
4	2000	.4665	933.00	.4514	902.80
			8099.60		7970.60

$$\text{I.R.R.} = 21 + \frac{99.60}{8099.60 - 7970.60} = 21.77\%$$

Project C

Year	Cash Flow	$(1+.2)^{-n}$	P.V.	$(1+.21)^{-n}$	P.V.
		r = .20		r = .21	
1	5000	.8333	4166.50	.8264	4132.00
2	4000	.6944	2777.60	.6830	2732.00
3	3000	.5787	1736.10	.5645	1693.50
4	3000	.4823	1446.90	.4665	1399.50
			10127.10		9957.00

$$\text{I.R.R.} = 20 + \frac{127.10}{10127.10 - 9957.00} = 20.75$$

Summary of I.R.R.

Project	I.R.R.	Ranking
A	20.89	2
B	21.77	1
C	20.75	3

Summary of Rankings

Project	Cost	Payback Period	N.P.V.	Profitability Index	I.R.R.
A	25000	3	1	1	2
B	8000	1	3	2	1
C	10000	2	2	3	3

Well, what do you make of that? Depending on the method you use each project is ranked last and each project is ranked second. The only thing that one can say with certainty is that no method of calculation we have used ranks project C first. It is generally inferior to the other two and should not be considered. The advice we would offer you is to consider carefully all the implications of the choice you make. Remember what each method tells you and what you lose when you do make a choice. You are unlikely to change company policy, but policies followed for years may be out of date and a periodic review of all current methods of evaluation could yield surprising conclusions.

Exercises to Chapter 5.

5.1 Given that the current rate of interest on deposits is 10% would you prefer to buy a television set for £200 cash now, or by making a deposit of £100 now and paying a second instalment of £120 in one years time.

Would your answer be different if the second instalment were to be fixed at
a) £115.
b) £108.

5.2 A piece of equipment costing £1000 has an expected life of 5 years. It is estimated that the cash flow resulting from the use of the machine will be £400 a year. The rate of return expected from capital of this type is 15%.
Calculate   a)   the pay back period
               b)   the net present value
               c)   the internal rate of return of the capital.

5.3 A firm is faced by two alternative investment plans. Plan A will cost £750 and plan B £950. Both plans involve the purchase of equipment the life of which is four years, and it is estimated that the cash flows during that time will be:-

	Year			
Plan	1	2	3	4
A	300	400	300	200
B	500	400	300	300

A minimum return of 20% on cost of capital is required. What advice would you give the firm?

5.4 a) Explain what is meant by
      i)   the internal rate of return and
      ii)   the net present value of an investment project.
    b)   In each case indicate how the concept can be used to determine the viability of an investment project.
    c)   Use the present value method to determine whether a firm should borrow money at 14% interest per annum to undertake a project costing £10,000 but yielding a return of £3000 a year for 6 years.

5.5 A firm can purchase a machine for £12,000 which will save £5,000 a year in labour costs. The machine has running costs of £1000 in the first year of operation rising by £500 a year thereafter. The machine has to be overhauled after four years at an additional cost of £500 and can be sold at the end of its useful life of seven years for £800.

Should the firm buy this machine if it is borrowing funds at a rate of interest of 9%?

5.6 A firm is offered four methods of paying for a piece of capital equipment, viz:
     i)   a single payment of £12,000 immediately
    ii)   no initial payment but £5,500 at the end of each years for three years
   iii)   a payment of £4,000 immediately plus payments of £3,000 at the start of each year of the next four years
   iv)   a payment of £7000 at the end of the year and a similar payment after a further four years.

The firm will meet any expenditure from a sum of £20,000 lodged in a bank account earning interest of 12% per annum.
Which method of payment would you recommend the firm to adopt?

5.7 The directors of Lambskin Ltd are considering three investment projects, Project A costs £3,000, Project B £3,500 and Project C £4,000.

The expected cash flows are

year	A	B	C
1	800	1000	600
2	1000	1200	1400
3	2000	1800	1800
4	1500	1800	2200

which project would you recommend to the directors?

5.8 A manufacturer has £5,600 available to purchase a new plant. His accountant recommends that equipment should not be installed unless it yields an internal rate of return of at least 12%. He expects to make £600 net profits in the first year of operation, £800 in the second year and £1,000 a year thereafter. The expected life of the plant is ten years and it has a scrap value of £500.

What would you advise?

5.9 Discuss the problems a manufacturer is likely to encounter if he is attempting to decide between different investment opportunities if he decides to use

    a)   the payback period criterion
    b)   net present value

which do you prefer and why?

5.10 The returns anticipated on a machine costing £850 are

    year 1   £300
          2   £400
          3   £300
          4   £200

The current rate of return on capital is expected to be 20%.

    a)   Is the investment viable?
    b)   If the machine has a scrap value of £200 at the end of its four year life does this affect the investment decision?
    c)   The firm is liable to pay tax at the rate of 40% of the previous years receipts. What is now the position
          i)   ignoring scrap values
          ii)   allowing for scrap values.
    d)   From your calculations what would you conclude about the effect of taxation on the investment decision?

Chapter Six

The Organisation of Data

The modern business world has a great hunger for facts and data. Well organised data improves our understanding of problems, and helps us to take decisions wisely. Badly organised data is little better than worthless. Unfortunately, you will all-too-often come across data that is not organised — most firms have filing cabinets full of data that someone intends to organise 'one day'. In this chapter we will suggest methods of how data can be organised for meaningful analysis — we will take our first steps in the rewarding (though often confusing) world of statistics.

Most people are vaguely aware that Statistics is concerned with figures in one way or another. Equally, we think, most people are rather distrustful of the statistics that they see quoted in the press or on television. We must admit that we ourselves have some sympathy for the housewife who is told on the news one evening that the cost of living has gone up by only 2% this month, and then finds in the shops next morning that everything she buys has, in fact, risen in price by between 5% and 10%. When this sort of thing happens it is no wonder that people get the impression that statistics can be made to prove anything. And yet — if our figures are accurate and the information is presented properly — how can this be so? We would like you to believe right from the start that no genuine statistician will ever deliberately misrepresent information or use it to mislead people. It can be done of course. In life many people are unscrupulous, and later in this course we will tell you how they misrepresent information, with the strict warning that *you* must never do it.

The great weakness of Statistics, is that to the man in the street who has never studied it, the methods used by statisticians are a closed book. We hope that as you work through this course your own personal book will be opened and that you will understand the dilemma in which our housewife finds herself.

But before we begin to think of the techniques you will use and the calculations you will perform, let us stop for a minute to consider the raw material you will be dealing with.

Suppose that the student union in your school or college wishes to obtain information about its members — their age, sex, home area, whether they live in a flat, or at home, or in lodgings and so on. How would the union secretary go about collecting this information? The most obvious way is for each student to be issued with a questionnaire, posing the relevant questions, and asking for it to be returned to the secretary's office. No doubt some of the forms will be incorrectly completed: some students may genuinely misunderstand the questions: some may refuse to answer certain questions which they regard as personal: doubtless some, in the fashion of the great petitions of the nineteenth century, will be signed by Queen Victoria or Karl Marx. Yet, with all its faults, this mass of completed questionaires is the basic raw material for the statistical report that the union secretary wishes to produce.

Raw material such as this, collected at first hand, in response to specific questions is known as *primary data*; its characteristics are that it is obtained directly for the purpose of the survey which is being undertaken, and is, as yet, unanalysed.

Now, if your union secretary is lucky, he may also be able to obtain a great deal of information from the College administration, who, using enrolment forms as their primary data, may already have produced for their own purposes a fair amount of statistical information about students. Such information will, of course, have been produced for college purposes and may not be exactly what the union wants: but it is often useful additional information. Such data, which has already been collected for another, and different purpose, we know as *secondary data*. Usually it is of less use than primary data since it has already been processed and the original questionnaire is unlikely to have asked all the questions you would like to have asked. But whether it is primary or secondary, there can be very little statistical information which was not at one time to be found only in a pile of completed forms or questionnaires. The main task of any writer on statistics is to explain what the statistician does with his raw data between collecting it and presenting his report. So let us go back to your union secretary.

It is obvious that no-one would sit down and write a report in the form of 'Mary Smith is 17, lives in Durham, and is in lodgings here; Susan Yeung comes from Singapore and is in lodgings here ...'

We might as well hand over the completed forms to anyone who is interested since all that this type of report does is detail the information which is already given in detail on the questionnaire.

We can get a clue about the next stage of the analysis if we ask ourselves what it is that the union really wants to know. Surely the sort of information that is really wanted is how many students are 16, how many are 17 and so on; what percentage of students live at home; what proportion of students come from overseas. It is not the individual we are interested in so much as total numbers in given categories. The categories in this investigation may be age, sex, type of residence, number of hours a week spent on study and so on. Within each category students will vary. Some are 16; others are 17; some live at home, others in lodgings. We call each of these categories a *variable* because within each category students will vary. So we may now say that we are interested in a number of variables such as age, and more specifically in the value we can assign to each student within the range of values over which the variable extends. We may find, for example, that when we consider the variable 'age', 267 students are aged 17, 164 are aged 18 and so on up to the eldest student. The numbers of students whom we can place at each value of the variable we will call the *frequency*, because it tells us how often we will come across a student with this particular characteristic (that is, aged 18, or doing 27 hours a week private study, or travelling more than 15 miles to college). Thus the first step we must take is to decide what aspects of student life we are interested in and count up how many students are found within each of these categories. In so doing we are simplifying our data — reducing it to a more manageable form. In the process some detail is lost. We no longer know how old Brenda Jones is; but if we are interested we still have her completed questionnaire. On the other hand we do know that 267 students are aged 17 as well as much other general information.

Once we have reached this stage we are in a position to summarise our results in the form of a table and our work begins to look more like that of a statistician. Probably as a first tentative step we would produce a simple table dealing with only one variable. It might appear like this:

Age of Students attending ABC College

Age (the Variable)	Number of Students (the Frequency)
17	267
18	164
19	96
20	74
21 and over	23
	624

There is nothing wrong with our producing 15 or 20 tables like this, each concerned with a single variable, but surely it is better for presentation purposes if we could produce a small number of compound tables each showing several variables at once. Thus we could construct a double table showing the two variables, age and sex of students at the same time.

We have constructed this table by listing one of our variables vertically (age) and the other horizontally (sex). There is no golden rule, but it generally looks better if we tabulate the variable with the greater number of values vertically and that with the smaller number of values horizontally. Notice too that we have totalled both the vertical and the horizontal columns and that this adds to our information. We not only have the age distribution of male students and of female students but also the age distribution of the entire student population, and the total number of male and female students.

Age	Number of Students		
	Male	Female	Total
17	151	116	267
18	98	66	164
19	70	26	96
20	52	22	74
21+	18	5	23
Total	389	235	624

You may of course still argue that the table is still concerned with only one variable, age, and that all we have is two age distributions. Let us then extend our table to consider three variables, age, sex and type of accommodation. Obviously now we must further subdivide either the horizontal or the vertical columns. Again it is a good general guide to say that we believe it better to subdivide the horizontal rather than the vertical columns. But in doing this the variable in the vertical column tends to become the more important. So we must consider which is the most important variable, and this often depends on what we are trying to show. Let us suppose that in this case we are aiming to show that the type of accommodation a student occupies depends on his or her age. In this case we will list the ages vertically and subdivide the horizontal columns. Our table may now appear like this:

Age	Number of Students						
	At Home		In Lodgings		In Flat		Total
	M	F	M	F	M	F	
17	112	92	16	20	23	4	267
18	64	42	24	16	10	8	164
19	31	12	28	7	11	7	96
20	8	4	16	10	28	8	74
21+	2	3	3	1	13	1	23
Total	217	153	87	54	85	28	—
	370		141		113		624

You will readily appreciate what a vast amount of information a table such as this can give us: the number of students who live at home, subdivided into male and female and classified according to age, as well as the same information for those who are living in lodgings or living in a flat. You can understand too how much more information could be incorporated if we subdivided further the horizontal axis as well as some subdivision of the vertical axis such as the area of origin followed in each case by the age range.

There is one problem — the more we subdivide, the more complicated our table becomes, and there comes a time when it is so difficult to read it and understand it that we find that clarity has been lost rather than gained. It is true that one treble table, such as the one above, is better than three single tables. It is equally true that if we are considering eight or nine variables, three treble tables are better than one very complex one. And if you are wondering why clarity is so important think again what we have been doing. We have collected primary data, simplified it and classified it, and are now trying to present it to our union executive in a readily digestible form. How much notice do you think the executive will take of us if they cannot understand what our tables are all about?

Just in case you are ever in the position of having to construct tables to present the raw material you have collected, there are several points you should bear in mind. Let us call them the 'Principles of Good Tabulation'.

(a) Every table should have a short explanatory title at the head. At the end you should put a note of the source of the information you have used, whether it is based on your own survey or secondary data.

(b) The unit of measurement should be clearly stated, and if necessary defined in a footnote. Not many people, for example, would know offhand what a 'Long Ton' is. In addition the heading to every column should be clearly shown.

(c) Use different rulings to break up a larger table – double lines or thicker lines add a great deal to the ease with which a table is understood.

(d) Whenever you feel it useful insert both column and row totals.

(e) If the volume of data is large, two or three simple tables are better than one cumbersome one.

(f) Before you start to draft a table be quite sure what you want it to show. Remember that although most people read from left to right, most people find it easier to absorb figures which are in columns rather than rows.

As with most things practice is the best way of learning, and these principles will soon become second nature after you have drafted a few tables for yourself.

You might well ask at this stage whether this is all there is in the subject of Statistics. If it were you would all end up with distinctions. But the most important part of the work is still to come. No statistician (or student) worth his salt is content with a mere list of figures. He now begins to ask questions, the most important of which is 'What do the figures tell me?' We now begin, to analyse the figures, and statistical techniques are largely methods of extracting the utmost possible information from the data we have available. We could, for example, calculate the average age of students living at home, and compare it with the average age of students living in flats to try to determine whether we are right in assuming that the younger student will tend to live at home and the older student tend to be a flat-dweller. We can do the same thing for both male and female students to see if they behave differently. Let us simply say that there are many questions that the statistician can ask even from the simple data we have used so far.

We said earlier that the most obvious way for the union secretary to collect his data was to issue a questionnaire to each and every student. The results of his enquiry would cover every single student in the college — it will refer to what statisticians call the *population* of students. Beware of this term population. In statistics it does not mean the number of people living in a particular area. What it implies is that we have examined or obtained information about every single member of a particular group we are investigating. Thus we can talk of a population of telegraph poles, a population of shaggy-haired dogs, a population of ball-bearings and so on.

But is there any need for us to examine the population of students attending the college? If we wish to save time and money can we not do as so many public opinion polls do and take a sample of students? We could issue the questionnaire to, say 60 or 70 students only, or perhaps to every tenth student, and so reduce our raw data considerably. The *sample results* we obtain can then be applied to the population of students: if 12% of the sample live at home, we will argue that about 12% of all students in the college live at home.

Now, you may well argue that this can lead to wildly inaccurate results; and if you consider some of the results of public opinion polls in recent years it is apparent that things can, and do, go wrong. The sample chosen may be too small; it may not be representative of the population; the error arising as a result may mislead us. At this stage we will merely point out that in taking a sample we are in good company: an extremely high percentage of government statistics such as the statistics of Household Expenditure are based on samples which, on the face of it, appear to be ludicrously small.

If you think back now to the questions we suggested that you might ask about what our tables can tell us, you will realise that most of them involve a more detailed study of one variable only — the age of females living at home; the age of males living in flats. When we do begin to analyse you will appreciate that this is usual. The table presents several variables at once, but we extract just one of them at a time for further examination. In a few cases we will use two variables at once, when we are asking if there is a relationship between them such that one affects the other or that both move in sympathy. But in this foundation course we will never ask you to get involved in the analysis of three or more variables at once — which is indeed a complex matter.

The Frequency Distribution

The problems we have met so far are simple. We have had to deal with only five age groups of students, and hence our tables are compact and easily presented. Let us now turn to far more complex data collected from the community as a whole. Government departments collect a mass of detailed information on which to base future policy and one of the areas which they investigate closely is the weekly earnings of adults aged 21 and over who are in full-time employment each April. In the United Kingdom it seems that earnings in the recent past ranged from about £10 a week to £80 a week and more. You can imagine how cumbersome a table we would have if we listed every single wage level as we listed each age. Theoretically we might have to list earnings at intervals of a penny, and so have a table of over 70,000 lines — a ridiculous situation. Fortunately we can simplify this. Do you think it really matters to us whether a worker is receiving £11.90 or £12.15 a week? We would probably be content to say that they both receive about £12 a week. Extending this further, we could argue that if we are looking at the pattern of wages we would be satisfied to combine together all those who received between £10 and £12 per week, or even between £10 and £15 per week. Now this is precisely what the government does in many of its published statistics. It does not say that 157 people receive £10.21 a week, 362 receive £10.37 and so on, but merely that 100,000 people receive over £10 but less than £15 per week. Earnings are stated in convenient bands or classes, and certainly we get a much better picture of earnings from this than we could ever get from a table listing 70,000 different levels of earnings.

To show you the effect of combining different values of the variable into a single class we reproduce below a table taken from the Annual Abstract of Statistics for the United Kingdom.

Distribution of Earnings of Adults in Full-time Employment

Earnings (£)		No. of Workers (millions)
10	and under 15	0·1
15	20	1·1
20	25	2·1
25	30	2·3
30	35	1·9
35	40	1·4
40	45	0·8
45	50	0·5
50	60	0·4
60	70	0·3
70	80	0·1
		11·0

Source: U.K. Annual Abstract of Statistics

A table such as this is known as a *frequency distribution* and is very common in statistical work. The variable we are considering, earnings, can take any value within the relevant group — it need not be an integer (or whole number). You might object of course that a worker cannot be paid £12.3762 a week, and in this sense it is not true to say that wages can take *any* value. On the other hand, subject to the minimum currency unit of

Britain being £0.005 or a half penny, it is a fair statement to make. This type of data is known as *continuous* data. Other examples of continuous data are ages, petrol consumption per week, miles travelled by British Rail, and you can probably think of dozens of others yourself.

We might, on the other hand, be considering a data variable which can have only set values. If we are considering the number of seats in a cinema there can only be a certain number, 875 or 358, not 364.34 or 786.93. Similarly, the number of journeys made by bus, the population of a city, the number of students sitting an examination, must be integers. Data of this kind which can have only a limited number of values is known as *discrete* data. An example of a frequency distribution using discrete data would be a frequency distribution showing the value of orders received by a departmental store during the winter months.

Orders Received

Value of Order (£000)	Number of Orders Received
0 and under 5	20
5 10	51
10 15	139
15 20	116
20 25	31
25 30	14
30 and over	5
	376

Few people would prefer to be faced with a table thousands of items long when they can have instead a manageable frequency distribution like these. Yet, something is lost. Most distributions are constructed, as was our table of wages, using groups or classes which may in some cases be very wide indeed. We have sacrificed detail for the sake of presenting a picture which can be absorbed fairly simply. It might seem, of course, that the use of class intervals will prevent our using the frequency distribution as the basis for further work. Naturally it does create a problem, and to overcome it we have to make an assumption. Going back to our wage data, we do not know the exact wages of the 1.9 million people earning between £30 and £35 a week. To enable further work to be carried out we assume that all the 1.9 million workers earn a wage precisely at the centre of the class we are considering, i.e.

$$\frac{30 + 34.99}{2} = 32.495 \text{ or } £32.50$$

Since you will be using this mid-point of a class many times, let us stress the importance of accuracy here. You must firstly consider whether the distribution is continuous or discrete. Consider a table showing cinema size determined by number of seats. One class may be cinemas ranging upwards from 200 seats but having less than 300 seats. Here the data is discrete; the minimum number of seats is 200 and the maximum is 299. The mid-point of this class is

$$\frac{200+299}{2} = 249.5$$

But if the same class, 200 and under 300, referred to continuous data, say the number of miles travelled by car, the minimum and maximum values of the group are 200 and 299.999. Hence the mid-point of the class becomes

$$\frac{200+299.999}{2} = 250$$

You will find, if you look at any published statistical tables, that in many cases no limits are given for the first and last classes. An income distribution showing annual income might begin merely with 'Under £660' and end up with the group 'Over £50,000'. Such open-ended classes create problems and we will give you a few hints on how to handle them later.

One of the most difficult problems you will have in building up a frequency distribution from raw data is to decide on what class intervals to use. Obviously, a great deal will depend on the data you have available, but a few general guidelines may help. Firstly, try not to choose class intervals which will reduce the number of groups below five or six. If you do the data will be so compressed that no pattern emerges. Naturally the rule is not infallible — E.E.C. have published statistics of farm sizes giving only three classes. These three, however, correspond to a generally accepted international definition of small, medium and large farms. Our advice is that you should try not to emulate E.E.C. Equally, at the other extreme, do not have too many classes. About fifteen or sixteen is the maximum. The problem here is not only the difficulty of absorbing lengthy tables, but also the fact that each group will have a very low, or in some cases, even a zero frequency. And this leads to another point. At the upper end of the table, if you stick slavishly to a single class interval you may well find that several consecutive groups have no members while a higher group has a frequency of two or three. In these circumstances you should sacrifice the idea of equal class widths and combine the several classes into a single wider class.

A good general guide is to take the difference between the minimum and maximum value of the variable (which we call the *range*), and divide by ten. This will give you the right class width (or thereabouts) for the majority of classes, provided that you realise that class width of five or ten, or fifty is better than one of four, or seven or sixty-two, and provided that you take care with the extreme values of the variable.

One final word of caution. Your table must be so constructed that each item of data falls clearly into a given class. Far too many tables have class intervals such as 10 – 20, 20 – 30, 30 – 40 and so on. But suppose an item is exactly 20 or exactly 30, which group do you place it in? If the distribution is continuous it is better to say, 10 and under 20, 20 and under 30; or

possibly 10 – 19.9, 20 – 29.9: if the table is discrete you can give the exact limits, 10 – 19, 20 – 29 and so on.
Before we leave this brief description of the frequency distribution it would be an advantage if we show you how to tackle examination questions which ask you to construct a frequency distribution from a mass of figures. For this purpose we will look at a typical examination question.

Example

The following is a record of the percentage marks gained by candidates in an examination:

65	57	57	55	20	54	52	49	58	52
86	39	50	48	83	71	66	54	51	27
30	44	34	78	36	63	67	55	40	56
63	75	55	15	96	51	54	52	53	42
50	25	85	27	75	40	37	46	42	86
16	45	12	79	50	46	46	59	57	50
56	74	50	68	52	61	40	38	57	31
35	93	54	26	67	62	51	52	54	61
93	84	28	66	62	57	45	43	47	33
45	25	77	80	91	67	53	55	51	36

Tabulate the marks in the form of a frequency distribution, grouping by suitable intervals.

Looking at these figures we find that there are 100 marks given ranging from 12 to 96. We have laid down a principle of aiming at somewhere in the region of 10 classes in our frequency distributions and it certainly seems that the best class width in this example would be 10 marks. If we were to use 5 mark intervals we would end up with some 18 classes which is too many; if we use 15 mark intervals we end up with only 6 classes which is too few.

We now have to find out how many of these marks fall within each class, and we recommend that you should do this in this way. Firstly list every class vertically; now take each candidate's marks in turn, and place a dash or a 1, or some other suitable mark against the class into which it falls. Having done this for every mark we can now take each class in turn and add up how many candidates fall into each class. Your rough working will appear something like this. You will notice that for ease of counting we have divided our dashes into groups of five.

10 and under 20	111	3
20 and under 30	ⅢⅠ 11	7
30 and under 40	ⅢⅠ ⅢⅠ	10
40 and under 50	ⅢⅠ ⅢⅠ ⅢⅠ 1	16
50 and under 60	ⅢⅠ ⅢⅠ ⅢⅠ ⅢⅠ ⅢⅠ ⅢⅠ 1111	34
60 and under 70	ⅢⅠ ⅢⅠ 111	13
70 and under 80	ⅢⅠ 11	7
80 and under 90	ⅢⅠ 1	6
90 and under 100	1111	4
		100

Before you do anything else now, check that the total frequency (that is, the number of dashes) in your rough working is the same as the number of items given in the question. Having done this you are now ready to construct your frequency distribution. Remember, though, all the things that are necessary: the heading of the table, the column headings and the source, if it is available. Getting the table correct is only one part of the answer, although it is an important part. Your final frequency distribution will appear something like this:

Percentage Marks gained by Examination Candidates
.... Examination .... 19 ..

Marks Awarded	Number of Candidates
10 and under 20	3
20 30	7
30 40	10
40 50	16
50 60	34
60 70	13
70 80	7
80 90	6
90 100	4
	100

Source: Examiners' Report 19 .

We will spend some time later on examining frequency distributions such as this to see what further information they can give us. But whatever distribution we study it has one thing in common with all other frequency distributions — it tells us the magnitude of a variable at a given point in time. There are times however when we need to look at the way the magnitude of a variable changes over fairly long periods of time — for example, we might be considering the way in which the volume of British exports to Hong Kong has changed year by year since 1960. Such a table is called a *time series*. It follows from the nature of the time series that it consists of a series of time periods: years, quarters, months, or even days, with the value of the variable given against each time period. Thus, recently, Barclays Bank conducted a survey of the output of motor vehicles quarter by quarter in each of the Common Market countries. The figures for France represent a typical time series with the value of the variable given quarterly.

Output of Motor Vehicles — France
(monthly averages — thousands)

Year 1	1st quarter	150
	2nd quarter	165
	3rd quarter	104
	4th quarter	113
Year 2	1st quarter	173
	2nd quarter	180
	3rd quarter	124
	4th quarter	184

Source: Barclays Bank Briefing No. 12

In much statistical work a time series of this nature may extend over many years, but you will find it surprising what such a series can tell us.

If you have followed carefully the argument of this chapter it should now be apparent that we are concerned with the collection, simplification, presentation and analysis of information which can be expressed quantitatively. If you do your work well you will get an accurate picture of the data you are studying. There will still be some of you who may claim that figures can be made to prove anything, and to be fair, if the data is misused this may be true. If you ignore parts of the data, and conveniently forget to include calculations and information which is inconvenient or does not support your preconceived ideas, most things are possible. But if you follow your analysis through to the bitter end, using all the information which is available, keeping an open mind and interpreting only what your figures throw up, you cannot but be somewhere near the truth. In a world, in which so much reliance is placed on the work of statisticians, where government policy often depends on their findings, and where industrial decisions involving millions of pounds are taken on the basis of statistical anaylsis, it is important to remember that whether you like what your figures show or not, your task is to interpret what the available data tells us as honestly as you are able.

Assignment Exercises

6.1 *Average Weekly Earnings* of Administrative, Technical and Clerical Staff in the Public Sector and in Insurance and Banking*

MALES

October	National and Local Government including Education (Teachers) and National Health Service	Nationalised Industries	Insurance and Banking
	£	£	£
1966	26.69	26.25	26.63
1967	27.88	27.13	27.73
1968	29.65	28.95	29.11
1969	32.03	31.18	30.88
1970	36.00	35.83	34.63

*Including earnings of monthly-paid employees converted to a weekly basis.

Source: Department of Employment.

Write a short report in which you bring out the main features of the data given above. Include appropriate derived statistics and illustrate your report with a suitable diagram or graph.

6.2 In 1951, 207 thousand persons received unemployment benefit, 906 thousand persons sickness benefit, 1437 thousand males retirement pensions, 2709 thousand females retirement pensions, 457 thousand received widows' benefit. 217 thousand persons received other National Insurance benefits. In 1971 the corresponding figures for unemployment benefit was 457 thousand, for sickness benefit was 969 thousand, for male retirement pensions 2611 thousand, for female retirement pension 5196 thousand, for widows' benefit 448 thousand, for other National Insurance benefits 387 thousand. (Source: *Social Trends* 1972.) Tabulate this data, calculate appropriate secondary statistics and include those statistics in your tabulation. Comment briefly on your tabulation.

6.3 An inquiry into the population of a town at 1st April 1974 showed that the total was 297,500 persons of whom 60% were females and 40% males. 50,000 females were aged thirty and under, 60,000 were aged from 31 to 60 and the remainder were over 60 years of age. The corresponding figures for males were 60,000, 39,000 and 20,000. The average family size was 3.5. 75% of the female population lived in the northern area of the town and the remainder in the southern area. Of the males, 20% lived in the southern area. Tabulate the data given showing the analysis of each class of person into areas and age groups. Include actual figures and percentages. Show also the number of families for the town in total.

6.4 The Saturn Finance Company wishes to study for several years (1971, 1972, 1973, 1974) the distribution of its loans according to size of loan (under £100, £100 and under £250, £250 and under £500, £500 and over) and the purpose of the loan (home improvement, car purchase, durable household goods purchase, other).

(a) Prepare a table in which the data can be presented cross-classified by year, size of loan and purpose of loan. Include summary rows and columns for all classifications given.

(b) Insert the following figures in the appropriate cells:

(1) In 1972, 42.7 per cent of all loans made for home improvements were for £100 and under £250.

(2) In 1971, 31.4 per cent of all loans made for purchase of cars were for £250 and under £500.

(3) In 1974, 29.7 per cent of all loans were under £100.

6.5 While the population of the United Kingdom grew from 38.2 millions in 1901 to 55.8 m in 1972, the increase was not uniform between the regions. Seventy years ago, the south-east was the most populous region with 10.5 m. In mid-1972, its population was an estimated 17.3 m. Both the West Midlands and East Midlands have also grown quickly; the former from 3.0 m to 5.1 m and the latter from 2.1 m to 3.4 m. During the same period, East Anglia's population grew from 1.1 m to 1.7 m, whilst the south-west increased from 2.6 m to 3.8 m.

Despite its generally high birth rate, Northern Ireland's population, whilst growing from 1.2 m in 1901 to 1.5 m in mid-1972, fell as a proportion of total U.K. population. A similar trend is revealed in other economically depressed regions, and both Wales and Scotland, because of losses due to migration, have grown much more slowly than average, Wales from 2.0 m to 2.7 m and Scotland from 4.5 m to 5.2 m. (Source: *New Society and Social Trends,* Modified.) *Note:* not all regions have been included.

Arrange the above data in a suitable table, providing additional columns, in blank, appropriately headed, in which could be inserted derived statistics enabling comparisons to be made.

6.6 The total number of employees of Core and Peel Ltd. at 31st December 1971 was 10,590, of which 6721 were men, 3106 women and the rest juniors. During 1971 108 men resigned and 74 men were engaged. The corresponding figures for women were 29 and 87 and for juniors 17 and 23. 1386 men, 976 women and 16 juniors were absent sometime during the year due to illness, 509 men, 876 women due to domestic circumstances and 366 men, 272 women and 3 juniors due to other causes. The average weekly wage rate paid to men was £32.00, to women £20.13 and to juniors £18.25. The company worked 50 weeks in the year. Tabulate these details showing suitable totals and sub-totals. Include also:

(a) the numbers employed at 1st January 1971,

(b) an estimate of the total annual wages paid per grade of labour and for the employees in total.

6.7 (a) The records of the Family Expenditure Survey of 1971 show that in 4642 households interviewed in the survey, the head of the household was in employment. In 1236 households the occupation of the head of the household was 'professional'; 101 earned less than £30 per week; 153 earned £30 but under £40 per week; 194 earned £40 but under £50 per week; the remainder earned £50 or over per week. In 470 households the occupation of the head of the household was 'clerical'. 119 earned less than £30 per week; 129 earned £30 but under £40 per week; 90 earned £40 but under £50 per week; the remainder earned £50 or over per week. In the remaining households the occupation of the head of the household was 'manual'. 930 earned less than £30 per week; 814 earned £30 but under £40 per week; 638 earned £40 but under £50 per week; the remainder earned £50 or over per week. Tabulate this data.

(b) Using your table as an example list the basic rules which should be observed when tabulating statistical data.

6.9 In 1970, 44,000 houses were purchased with a local authority mortgage (total sum advanced = £154m) and 32,000 with the help of insurance companies (£154m advanced) whilst building societies lent £1986 m to 540,000 purchasers. These figures compare with 56,000 (£168 m), 34,000 (£124m) and 504,000 (£1,477m) in 1967, and 19,000, 40,000 and 460,000 houses purchased in 1969 with the help of loans of £69m, £179m, and £1556 m from local authorities, insurance companies and building societies respectively (abridged from *New Society*).

(a) Arrange the above data in concise tabular form.

(b) Prepare a table in blank, with suitable headings in which could be inserted derived statistics from the above figures which would facilitate the making of comparisons.

6.9 (a) List clearly the steps taken in forming a frequency distribution from a set of about 1000 observations giving reasons where necessary.

(b) Construct a frequency distribution using the following 100 observations:

Lives of electric light bulbs, in hours, to the last complete hour.

690	701	722	684	662	699	715	742	726	716
728	705	693	691	688	706	707	691	701	713
740	662	676	738	714	703	695	692	699	685
698	687	703	726	699	692	714	724	664	689
694	705	717	682	717	707	696	697	681	708
712	733	705	673	694	716	745	692	719	701
679	680	654	691	669	685	725	704	724	714
689	702	710	696	697	709	721	677	680	671
685	724	736	696	688	692	728	656	690	695
702	696	708	698	710	682	694	676	700	663

6.10 The following is a record of the heights in centimetres of a sample of 85 servicemen:

169 179 183 186 166 181 177 173 167 193 176 183 162 170 186 174
188 165 168 174 170 176 186 177 185 175 179 166 190 182 182 180
194 177 184 175 168 181 180 172 178 192 175 189 180 175 183 191
172 188 180 176 185 178 179 173 165 170 178 181 181 189 187 191
179 196 179 182 171 169 171 184 198 182 175 190 187 176 164 187
167 185 177 184 178

Tabulate the above data in the form of a frequency distribution, using as intervals 160 cm and under 165 cm, 165 cm and under 170 cm, 170 cm and under 175 cm, and so on. Illustrate the table by means of a histogram.

6.11 The lengths of telephone calls from a certain office were noted and the results are shown below giving the times in seconds.

141	43	203	104	82	63	24	84	41	86	47	43
100	53	139	147	137	186	214	106	150	109	170	172
194	124	175	177	162	129	128	219	40	105	48	65
105	154	154	35	149	54	104	109	119	74	140	104
168	127	191	30	109	88	104	207	38	164	182	120
166	53	145	29	112	143	49	199	130	52	109	77
142	75	146	105	125	112	40	126	67	49	90	140
132	118	134	133	159	123	161	112	157	104	92	112
151	98	156	117	156	190	122	135	116	96	163	116
186	155	106	153	69	105	136	106	131	118	94	121

Arrange these figures in a grouped frequency distribution using the intervals 0 – 19, 20 – 39, etc.

Chapter Seven

The Presentation of Data I

Numerical data, however well organised, can often fail to fulfil one of its prime functions — to communicate information. This is because many people have a positive hatred of numbers. Be honest — how often have you read a newspaper article and simply skipped over any tables it contains? It is not enough to organise data well: it must also be well presented. One of the main jobs of a statistician is to identify the main features of information given by data, and to present them in such a way that they become intelligible and interesting. In this chapter, we will examine some of the graphs and diagrams that the statistician uses to achieve this end.

Watch any television education programme and almost certainly one of the first things to strike you will be the number of devices used to present information in a vivid and interesting fashion. Graphs, diagrams, blocks of wood, animated cartoons — all play their part in putting the subject over. Of course, television is the ideal medium for this type of visual presentation, and one of the more important developments of the second half of the twentieth century may well turn out to be the impact of television as a means of imparting information. Yet, in all this, the television producer is doing little more than statisticians have already been doing for a considerable period of time. The scale is larger; the impact is probably many times greater; but the techniques are the same.

We would like, at this point, to be able to say to you that we are going to give you a few simple rules which will enable you to master the technique of presenting diagrammatic information quickly and easily. It is not, however, as simple as that. The diagram you draw depends in part on the information you have, and what you are trying to stress. If you are trying to stress how much the government is spending annually you will produce a very different diagram from one which sets outs to examine what the government spends the money on. It might also depend on the readers you are aiming at. A diagram suitable for the readers of a mass circulation daily newspaper will be very different from one in an accounting journal; a diagram intended to extol the qualities of a particular brand of soap powder will be far removed from one designed to show the changing composition of agricultural output in the Common Market. Effective presentation is a question of flair and experience, and there is only one guideline — does the diagram present clearly and vividly the information it is designed to present?

We cannot pretend that the way in which we illustrate the presentation of particular information is the only way. You may, in fact, think it is not the best way. What we will do is to indicate the weapons you have available. The rest is up to you!

1. The Bar Chart

One of the most common of all techniques for presenting data is the use of the bar chart, in which the length of the bar is proportional to the size of the items we are considering. Suppose we are considering the population of some E.E.C. countries; we could be presented with a table like this:

	Population (million)		Population (million)
France	50	Denmark	4.9
Belgium	10	Britain	56
Germany	61	Holland	12.9
Italy	54		

Now, shut your eyes, and without reference to the table state which country has the highest population, which is the lowest, and where Britain comes in the league table. If you have played fair we are prepared to bet that a high proportion of you cannot answer all three questions correctly. Why? Simply because experience shows that people do not absorb lists of figures. A very different result would be obtained if we presented the same information in the form of a bar chart.

The immediate impression we get is of four giant members accompanied by a number of very small members; and it hits the eye that Britain is a giant, second only to Germany.

You will notice that the bar diagram has a heading, as should all diagrams. It is no use presenting information unless the reader knows what the information is. Note, too, that the source of the information is given as we are using secondary data. This enables the reader, if he is interested, to go back to the original figures and delve more deeply.

Here we were looking at one variable factor at a particular point in time, but we may equally use the bar diagram to show how the value of something, say the output of motor vehicles in France, has varied over time.

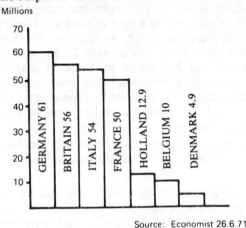

Source: Economist 26.6.71

Population of E.E.C. countries 1969 (millions)

Diagram 7.01.

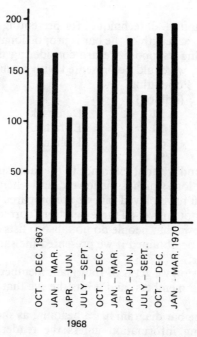

Source: Barclays Bank 3.10.70

Production of motor vehicles — France (monthy average – thousands)

Diagram 7.02

Here you will notice that the bars are separated — this is a matter of personal preference. It is probably neater, but many statisticians say that it is more difficult to compare the height of the bars. Looking at the diagram it is immediately apparent that production in three quarters was exceptionally low; in April – June of year 2 and in the July – September quarter in years 2 and 3. Further examination may show that in the July – September quarter production is low every year, possibly due to its being the holiday season, but the April – June quarter of year 1 is clearly exceptional and one is constrained to make a more detailed examination of what happened at that time.

The bar chart is a most versatile instrument and capable of adaptation to almost any data. Three further modifications will, we think, convince you of this versatility.

Often the data we are examining includes negative figures. Profits in one year, for example, might be converted into losses the next; the balance of payments may be in surplus or deficit. Such negative figures can be represented on the bar chart quite simply by extending the bar below the zero base line as in Diagram 7.03.

Sometimes, of course, we may be less interested in an absolute total than we are in the way that total is made up. We may wish to find out how the final cost of production is made up, where sales revenue goes to, what the government does with the money it collects. Again the bar chart proves itself equal to the task, since we can always subdivide the bar as shown in Diagram 7.04.

Such a bar chart is often called a compound or component bar chart, since it illustrates the components that go to make a total. Sometimes we may be more interested in expressing our information in the form of percentages. It may be important to our argument that tax on North Sea Oil is 54.4% of total selling price rather than that it is 6.8 dollars. There is nothing to stop us from constructing a component bar chart the length of which represents 100% subdivided into sections to show the percentage that each component item forms of the total. In the case of tax, then, the section of the bar representing tax paid would be just over one half the length of the bar. Such a diagram, for obvious reasons, is called a percentage component bar chart. As a final example of the use of the bar chart we will try to show how it can be used to derive information which is not immediately obvious. With Britain's entry into E.E.C. a burning question is the efficiency of European agriculture. But how can efficiency be measured? It is fair to say that if 30% of the total labour force is engaged in agriculture but the agrarian output is 10% of the Gross National Product, that country's agriculture is inefficient compared with a country for which the figures are 10% and 12% respectively, and this will be the criterion we will use.

We think you will agree that, using our criterion of efficiency, of the seven countries shown, only three — Holland, Belgium and Britain — can be said to be efficient producers in the agricultural sector, and this may well be a factor in the problems facing the common agricultural policy of the E.E.C.

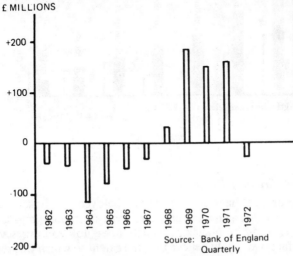

Diagram 7.03   **U.K. net invisible earnings from non-sterling areas**

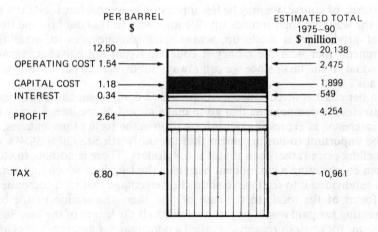

Costs of North Sea oil

Diagram 7.04

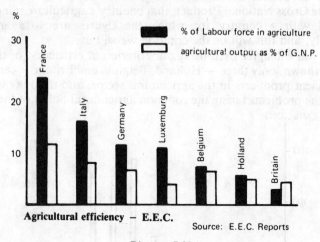

Agricultural efficiency — E.E.C.

Source: E.E.C. Reports

Diagram 7.05

2. *The Pie (or Circular) Diagram*

The circular, or pie diagram is a device beloved of those who present statistical data for the general public. It is a rare report, and a still rarer copy of the *Economist* magazine which does not contain several such diagrams. In fact, we firmly believe that the extent to which this technique is used grossly exaggerates its utility.

This diagram has only one real use — to show the relative size of the component parts of a total. A complete circle represents the total and this circle is divided into segments the size of which represents the relative importance of each constituent of the total. Thus, if we were trying to show the nature of road accidents in a particular area, we might find that of 300 accidents occurring last year, 57 involved motor-cycles. We now have to mark off a segment of the circle corresponding in size to the proportion of accidents involving motorcycles, that is, 57/300 or 19/100. Since there are 360 degrees in a circle the appropriate segment must subtend an angle of 19/100 × 360 degrees at the centre, i.e. an angle of 68.4 degrees. Our complete pie diagram will appear something like this:

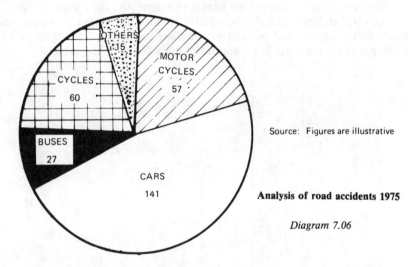

Source: Figures are illustrative

Analysis of road accidents 1975

Diagram 7.06

Even with this single simple example you should find it easy to spot the weaknesses of the pie diagram:

(a) It involves cumbersome calculations.

(b) It is more difficult to compare segments of a circle in a pie diagram than to compare heights in a bar diagram.

(c) It gives no information as to absolute magnitude unless figures are inserted in each segment, whereas the bar diagram is scaled against a single axis.

Our own advice is not to use this method of presentation unless it is forced upon you.

3. *Plotting the Frequency Distribution — the Histogram*

You will remember that the frequency distribution examines the frequency of occurence of different values of a variable at a given point in time, and that the values of the variable are combined together into classes of a predetermined size. It is this type of distribution that the histogram presents. The histogram is so similar to a bar chart that students often assume that they are one and the same thing. There is, however, a major difference between them. In a bar chart we are interested in only one factor, say output, the magnitude of which can be represented by the height of a bar. It does not matter how wide the bar is; we look only at the height. In a histogram, however, we are interested in two factors – the width of the classes represented by the width of the bar, and the frequency with which items within each class are found, represented by the height of the bar. In these circumstances it is the area of the bar which really interests us. Now, admittedly, in many of the cases you will deal with this distinction makes little difference. The class intervals are the same throughout the frequency distribution, so the width of the bars is the same and the relative heights of the bars will indicate directly the relative frequencies. Often, however, class intervals at the top and bottom of the distribution are left 'open', or are different from the class intervals in the main body of the table.

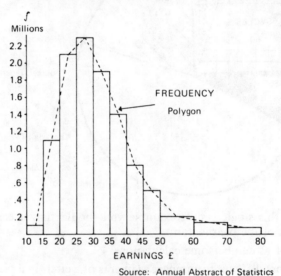

Source: Annual Abstract of Statistics

Weekly earnings of full-time adults 1971

Diagram 7.07

Suppose, for example, that the last group is twice as wide as every other class in the distribution. We will have to double the width of the bar to represent this doubling of the class interval. Remember though, it is the area of the bar that we are interested in, and so, having doubled the width, we will have to halve the height. This is not as confusing as most students believe if you think about what we are doing. In doubling the class intervals we are combining into one wider group items that belong to two of the narrower groups. To be consistent, then, we have to make an arbitrary assumption that the frequency of the wider class is divided equally between each of the narrower classes. Suppose that throughout our table the width of the class is £5, but the width of the last class is increased to £10. It appears as '£80 and under £90'. In this class appear items which we would have expected to appear in the two classes '£80 and under £85' and '£85 and under £90'. Doubling the width of the bar indicates that we are combining these two classes into one, and to illustrate the frequency we assume that each of these two 'narrow' classes contains half the number of items in the 'wider' class. That is, we halve the height of the wider bar. Be careful, however, not to draw a line down the middle of the wider bar to try to indicate this. We are only making an assumption — if you do draw such a line you are saying that if we plotted the two 'narrow' groups separately each would have exactly the same frequency, and this we cannot know.

Bearing this in mind, let us plot the histogram of the frequency distribution of weekly earnings that we obtained in the last chapter. (Diagram 7.07).

Note especially the last three groups where the class interval rises from £5 to £10 and the height of the bars is halved.

A similar picture of the distribution is given by the frequency polygon which converts the histogram into a simple graph. Notice that we form the graph by joining the mid-points of the top of the bars. You will find that this convention of using the central value of the group to represent the group as a whole is a very common technique in statistical analysis. Here we are joining the mid-points by straight lines to give us the polygon, but sometimes you will be asked to construct a frequency curve and will join the mid-points of the top of the bars by a smooth curve instead of straight lines. Frequency curves of this type are very common when we wish to illustrate particular types of frequency distribution.

Finally, it is often useful to present the distribution in a different way altogether. Instead of tabulating the frequency of each class we would like to know the frequency with which the variable falls *below* a particular value. The Chancellor who wishes to help the more poorly paid in society might well propose to introduce a tax exemption bill applying to all those earning less than £35 a week. Naturally, he will wish to know firstly how many people this bill will affect, and, secondly, how many more wage earners than before will now be free of the burden of income tax.

It is, of course, simple to add up the frequencies of the first five classes of our frequency distribution, but there is more than this. For many purposes it is easier to read off directly the frequency we require, and there are measures we will introduce you to in a later chapter which are far, far easier to obtain graphically than to calculate.

To obtain information of this nature the statistician has devised the Ogive or Cumulative Frequency graph. This diagram plots, on the horizontal axis, certain values of the variable, usually the upper value of each group; on the vertical axis it shows the frequency of the items with a value less than this.

Thus, in constructing the ogive, we first construct a cumulative frequency table from our frequency distribution in this way:

Wage (£)	Cumulative Frequency (millions)
Under 15	0.1
20	1.2 (i.e. 0.1 + 1.1)
25	3.3 (i.e. 0.1 + 1.1 + 2.1)
30	5.6

and so on until we reach the final group

| Under 80 | 11.0 |

It is this table that we plot as the ogive, in Diagram 7.08.

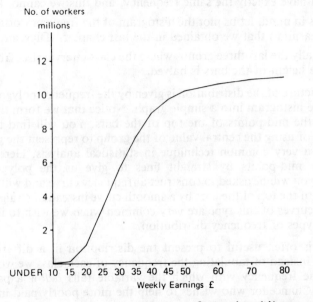

Source: Annual Abstract

Earnings of full-time adults April 1971

Diagram 7.08

The first thing you will notice is that we have joined the points on the ogive by straight lines rather than the smooth curve that you are probably more used to. The reason for this is that we do not know how the items in any group are scattered between the upper and lower limits; so we make the only reasonable assumption that we can − that the items in any group are distributed evenly across the group. It may be a false assumption. We may find, in the group £25 and under £30, that all the 2.3 million members get a wage of £29.50; but it is not likely. So long as there is a reasonable number of members within the group, we are fairly safe in assuming an even distribution. We know that it will not materially affect our results.

4. Plotting the Time Series

You will remember from the last chapter that much data is given in the form of a time series, in which we take a variable and show how its magnitude has varied over a period of time. We have obtained from the United Kingdom Annual Abstract of Statistics the following table showing how consumer expenditure at constant prices varied during the period of time when prices in the United Kingdom were rising at an alarming rate, 1964 to 1974.

Consumer Expenditure at Constant Prices
(£ million)

1964	28330	1970	31472
1965	28760	1971	32397
1966	29301	1972	34318
1967	29869	1973	35962
1968	30598	1974	35741
1969	30715		

Source: Annual Abstract of Statistics

Perhaps you will agree that these figures, involving thousands of millions of pounds, mean very little; if we are honest probably very few of you even read the figures in detail.

But suppose we now draw a graph of these figures!

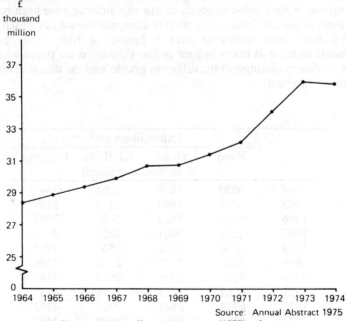

Source: Annual Abstract 1975
Consumer expenditure at constant (1970) prices
Diagram 7.09

We are sure that even a quick glance at this graph will leave you with a permanent impression of the way in which consumer expenditure has been rising – slowly and steadily from 1964 to 1970, then much more rapidly, reaching a peak in 1973. Since we have plotted expenditure at constant prices, this represents a rising consumption of goods, i.e. a rising standard of living (it may of course be at the expense of past saving).

You will have noticed the break in the graph on the vertical axis and the sudden jump from 0 to 25. This is a device used by statisticians when every figure is high and we do not want to crush our graph into a small space at the top of the graph paper. Remember too that we could have left you with a very different impression of the behaviour of consumer expenditure by adjusting the scales on the axis. Try it for yourself by doubling the scale on the vertical axis and halving the horizontal scale; or halving the vertical scale and doubling the horizontal. We have altered the scales in diagrams 7.10 and 7.11. After looking at three graphs of consumer expenditure can you really draw any conclusions as to the rate at which consumer expenditure has been rising? We know it has gone up – and that is about all.

Now, this is all very well, but our friends, the economists, will immediately want to know what we have been spending more on. Are we buying more food and clothes, buying more houses, or wasting our resources in riotous living? For their sakes, and for yours, we add below the way in which consumption of four sub-categories of our expenditure have been behaving in the same period. Whenever we wish to compare the way in which several variables have been behaving over a period of time, it is perfectly permissable to draw as many as four or five time series on the same axes, as long as we clearly distinguish the different graphs and the diagram is not too difficult to interpret.

| | Expenditure on | | | |
| | Food | Drink | Clothing | Housing |
		(£ thousand million)		
1964	6080	1866	2366	3481
1965	6081	1849	2426	3597
1966	6170	1922	2425	3695
1967	6228	2001	2450	3836
1968	6260	2108	2568	3973
1969	6264	2149	2596	4106
1970	6365	2296	2693	4181
1971	6362	2463	2726	4287
1972	6320	2641	2910	4377
1973	6388	2988	3060	4483
1974	6418	3081	3039	4460

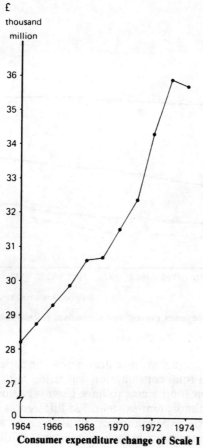

Diagram 7.10 — Consumer expenditure change of Scale I

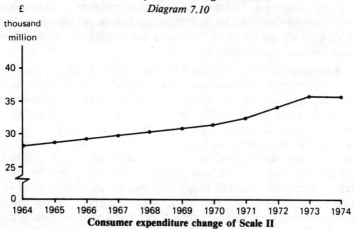

Diagram 7.11 — Consumer expenditure change of Scale II

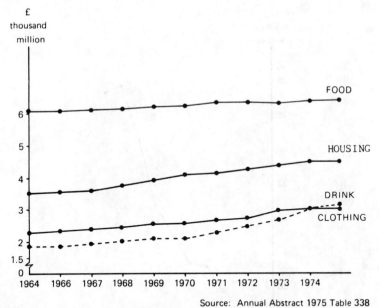

Source: Annual Abstract 1975 Table 338

Consumer expenditure at constant prices 1964-74

Diagram 7.12

You can see from the graphs we have drawn how the pattern of expenditure is changing. Although total consumption was rising fairly rapidly over this period, expenditure on food seems to have risen very slowly indeed — the slope of the graph is almost non-existent. Far different is the case of drink. Particularly from 1969 expenditure on drink rose at an alarming rate. Remember that this expenditure is listed at constant (1970) prices, so the rise of the graph represents increasing real consumption. Economists and sociologists would find this an interesting comment on human behaviour, and, in looking at this diagram, use your knowledge of other subjects to interpret what has been happening.

Do you remember when we discussed the bar chart we showed how it could be used to illustrate the constituents of a given total? Now, there are times when we want to illustrate how these constituents have varied over time. It may be, for example, that over the course of years road accidents involving bicycles have been forming a smaller and smaller part of total road accidents, while accidents involving motor-cars have been gradually forming an increasing proportion of total accidents. There is nothing to stop us from drawing three or four different time series, one for each type of vehicle showing the number of accidents involving that type of vehicle. We could even draw a number of graphs showing the percentage of accidents involving each type of vehicle, and we have no doubt that the diagrams would bring out the changing pattern. For this type of analysis it is far better to use a special type of diagram, the *Strata Graph,* or, if you are dealing in percentages, the *Percentage Strata Graph.* We must, however, keep the number of constituents reasonably few since otherwise the diagram may become difficult to interpret. The great advantage of this type of presentation is, as you will see, that the lines showing the magnitude of each consistutent can never cross. Diagram 7.13 is a typical strata graph, showing the constitution of road accidents over a period of time in a large city. In order to stress the points that we would like to bring out, the figures are purely imaginary, so do not think that your own town is abnormal if the pattern of accidents is different from the one we have represented.

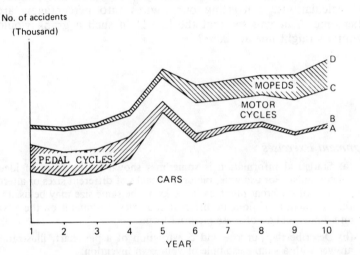

Number of road accidents 19... to 19... analysed by type of vehicle

Diagram 7.13

The principle of the strata graph is that our totals are successively cumulated. Thus the height of line A represents the number of accidents involving motor-cars each year. When we consider those accidents involving bicycles we add together for each year those accidents involving motor-cars and those involving bicycles to produce line B, the height of which for any year tells us the number of accidents involving (motor-cars + bicycles). In the same way line C is the graph of the number of accidents involving (motor cars + bicycles + motor-cycles). Finally when we add to these figures those accidents involving mopeds, we arrive at line D, which is the total number of accidents for all vehicles. You will have realised already that the distance between each line represents those accidents involving a particular vehicle. Thus the gap between A and B shows us accidents involving bicycles and the gap between B and C those involving motor-cycles. It is, of course, desirable to distinguish each type of vehicle by distinctive shading to give the diagram clarity, but we think you will agree that such a diagram is relatively easy to interpret. Even a cursory glance tells us that the total number of road accidents has been on the increase, but within this total three distinct trends are apparent. Bicycle accidents, which were quite numerous in the early years, have dwindled away and are now very few: conversely, moped accidents, which were very few in the early years, probably because there were few moped owners, are now becoming quite an important constituent of total accidents: car accidents have increased, and although it is difficult to judge, probably remain much the same proportion of an increased total. Motor-cycle accidents, it is apparent, have remained fairly constant over the years.

If we wish to draw a percentage strata graph, it is just as easy, although the initial calculations, converting our figures into percentages, are very cumbersome. Can you see that the line D in such a graph would be a horizontal straight line at 100%?

Assignment Exercises

7.1 (a) Statistical information is sometimes shown in the form of ideographs (pictograms). For example, pictures of sacks of different sizes or alternatively pictures of different numbers of sacks of the same size may be used to show the production of flour in different countries. Comment on the advantages and disadvantages of these pictorial forms.

(b) Describe the purpose and construction of a pie chart, illustrating your answer with a simple example of your own invention.

Scot. A.S.C.

7.2 Explain the use and method of construction of the following ways of representing data, illustrating your remarks with an example in each case:

(a) Bar charts,

(b) Pie charts,

(c) Pictogram.

7.3 (a) What are the advantages of using charts and graphs in statistical investigations?

(b) Describe clearly the methods to be employed in constructing:

(i) a pie chart,

(ii) a histogram.

(c) State the kind of diagram you consider most suitable to illustrate:

(i) daily hours of sunshine for a period of one month.

(ii) the number of workers, male and female, employed in a factory at each of three dates.

(iii) monthly sales, cumulative sales and a curve showing trend.

Briefly give your reasons for your choice in each case.

7.4 Discuss the suitability of a compound bar chart, a pie chart, and a pictogram for the presentation of data to:
   (a) top management,
   (b) lower management,
   (c) the public.

   Give examples of data which could be satisfactorily presented by each of these charts. (Do not draw the charts.)

7.5 A daily count of the number of rejects from the assembly line of a local manufacturer has yielded the following data:

138	164	150	132	144	125	149	157
146	158	140	147	136	148	152	144
168	126	138	176	163	119	154	165
146	173	142	147	135	153	140	135
161	145	135	142	150	156	145	128

   (a) Using the data, construct a frequency distribution table and from that sketch the corresponding frequency curve.
   (b) Comment on the shape of the frequency curve you have obtained and compare it with the sketched shapes of two others with which you are familiar.

7.6 The set of figures below shows the ages at which 50 employees were appointed to a certain grade.

28	27	30	27	28	28	26	27	28	28
26	28	29	31	27	28	27	29	27	29
28	27	31	27	27	29	30	27	28	28
28	29	28	29	27	30	27	28	27	29
28	27	28	29	29	28	28	28	28	27

   Write the data in the form of a frequency table and draw the frequency curve; by observing the shape of this curve, comment on the skewness of the distribution.

   Draw also the cumulative relative frequency curve.

7.7 Estimates of Gross Domestic Product (GDP) in current prices from 1951 to 1972 in £m are given below.

	£m		£m
1951	12,639	1962	25,279
2	13,790	3	26,878
3	14,877	4	29,187
4	15,726	5	31,156
5	16,867	6	33,057
6	18,264	7	34,835
7	19,369	8	37,263
8	20,196	9	39,667
9	21,248	1970	43,303
1960	22,633	1	48,675
1	24,213	2	53,940

Source: National Income and Expenditure, 1973

(a) Construct a frequency distribution and a histogram of these figures and comment.

(b) Indicate how these current price estimates can be converted to real (i.e. constant price) estimates of GDP.
C.I.P.F.A.

7.8 The following is a record of marks scored by candidates in an examination:

77	59	84	73	51	43	50	81	61	53	69
37	58	63	67	61	90	61	50	60	84	56
77	57	42	43	41	49	37	21	24	35	34
50	11	52	30	16	33	67	87	64	47	59
37	92	88	30	38	22	22	49	46	50	64
23	73	73	48	26	36	51	85	71	57	45

(a) Tabulate the marks in the form of a frequency distribution, grouping by suitable intervals.

(b) Construct a histogram from your frequency distribution.

(c) Explain the essential differences between a histogram and a bar chart.
Scot. A.S.C.

7.9 *Stocks of Coal – Great Britain, July 1972 to June 1973*
(thousand tons)

	Total	Opencast Sites and Central Stocking Grounds	Collieries
1972			
July	8,839	3,419	5,420
August	9,282	3,530	5,752
September	9,764	3,528	6,236
October	10,030	3,473	6,557
November	10,471	3,458	7,013
December	10,934	3,376	7,558
1973			
January	11,130	3,201	7,929
February	11,455	3,224	8,231
March	11,972	3,244	8,728
April	12,470	3,268	9,202
May	12,925	3,332	9,593
June	13,292	3,417	9,875

Source: Department of Trade and Industry.

(a) Write a short report stressing the main features of the data given above and include any derived statistics which may be appropriate.

(b) Prepare a suitable diagram or graph which will illustrate your report.

7.10 *Household Expenditure in 1971*

	£	Per Cent of Total
Housing	3.98	12.8
Fuel, light and power	1.85	6.0
Food	8.02	25.9
Alcoholic drink	1.46	4.7
Tobacco	1.30	4.2
Clothing and footwear	2.81	9.0
Durable household goods	2.01	6.5
Other goods	2.32	7.5
Transport and vehicles	4.26	13.7
Services	2.90	9.4
Miscellaneous	0.09	0.3
Total weekly household expenditure	30.99	100.0

Draw carefully and neatly a chart, graph or diagram to represent in visual form the household expenditure in units of actual money spent. Draw carefully and neatly a second chart, graph or diagram to represent in visual form the household expenditure in percentage terms. State the reasons for your choice.

7.11 *Unemployed in Great Britain Receiving Unemployment Benefit (thousands)*

	Total	Unemployment Benefit Only	Unemployment Benefit and Supplementary Allowance
1970			
February	332	260	72
May	303	238	65
August	286	226	60
November	305	245	60
1971			
February	401	312	89
May	406	310	96
August	427	321	106
November	494	379	115
1972			
February	514	391	123
May	451	339	112
August	385	291	94
November	344	261	83

Source: Department of Employment.

(a) Prepare a suitable graph or chart to represent the data given above.

(b) Write a short report on the main features revealed by the table and your graph or chart using derived figures where appropriate.

7.12

*National Insurance — Great Britain: New Claims —
Weekly Averages
(thousands)*

Year	Sickness and Invalidity Benefits	Injury Benefit	Unemployment Benefit
1966	206.1	18.5	50.6
1967	193.2	18.8	63.0
1968	204.0	18.0	58.8
1969	219.4	17.9	59.6
1970	204.5	15.8	60.7
1971	169.3	14.0	68.7

Source: Department of Health and Social Security.

Prepare a suitable graph or chart to show the data given above relating to National Insurance claims in Great Britain. Write a short report on the main points revealed by your graph, using derived figures where appropriate. Explain carefully what is meant by 'Weekly Averages' and why this term is used rather than that relating to any other time period.

Chapter Eight

The Presentation of Data II

The diagrams that we have examined so far are in fairly common usage. In fact we would be prepared to bet that, with the exception of strata graphs, you will have already seen all of the diagrams in the previous chapter. Basically, all of these diagrams are attempting to show the relative quantities in the data in an interesting and easily grasped manner. Yet some diagrams show much more than just the relative sizes, and we will now turn our attention to such diagrams. The chapter is concluded with a discussion on how diagrams can be made to misrepresent data.

1. *Logarithmic Graphs*

As we have seen, one of the problems with ordinary graphs is that they tend to give a false impression of the way in which figures are changing. Using a vertical scale of 1 inch = 100 units, a change from 100 to 200 units is represented by the same upward movement as a change from 1000 to 1100 units, even though it is 100% change as compared with a 10% change. The slope of the graph is the same − yet the rate of change is very different. Now, in most cases this does not matter. We may be interested only in the swings of the absolute figures; or the range of the figures may be such that the difference in the rates of change is negligible. If this is the case, we can get all we need from the graphs we have drawn. However, if we are considering inflation, for example, it is the rate at which prices are rising which is important, rather than the actual price increases.

Fortunately, if we plot the logarithms of our figures rather than the figures themselves, we can produce a graph the slope of which represents the real rate of change. Why should this be so?

Consider a case in which prices are rising by 20% a year. We can easily construct a table to show what happens to prices.

Period	Price	Log of Price	Difference of Logs
1	100	2.0000	
2	120	2.0792	0.0792
3	144	2.1584	0.0792
4	172.8	2.2375	0.0792
5	207.36	2.3167	0.0792

Don't you find it rather frightening that an inflation rate less than that of Britain in the mid 1970's will more than double prices in five years?

As you can see, the log of each prices rises by 0.0792 irrespective of the magnitude of the price change, and hence we obtain a straight-line graph if we plot the logarithms. This of course tells us that the rate of change of prices is constant. It does not tell us what the rate of change is, but merely enables us to compare rates of change over time.

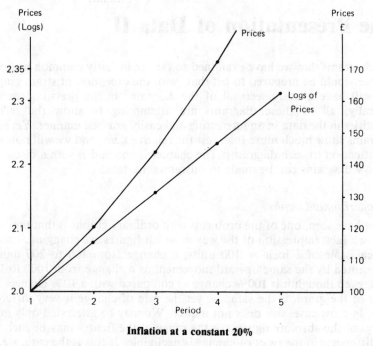

Inflation at a constant 20%

Diagram 8.01.

Strictly such a graph is known as a *semi-logarithmic* graph, as we plot logarithms only on the vertical axis. On the horizontal axis we do not plot the logarithms of the time periods.

As you can see from the diagram, if we plot the prices themselves the slope of the curve gets steeper and the fact that the rate of change is the same is completely hidden.

Often a logarithmic graph can be used also when the figures with which we are concerned range so widely that it is inconvenient to use a normal graph. Such a situation would occur if we tried to plot German price movements during the great inflation of the 1920's when prices were increasing by hundreds of times each month.

Now, although few students find difficulty in mastering the principles involved in the construction of semi-logarithmic graphs, this is such a common question in examinations that we would strongly recommend you to work through the following example with us, making sure that you understand each step.

Example

Sales of Two Companies
Sales (£000)

Year	Company A	Company B
1961	2240	980
1962	2460	1082
1963	2680	1205
1964	2915	1289
1965	3136	1382
1966	3362	1476
1967	3590	1580
1968	3821	1687
1969	4049	1787
1970	4280	1891

(a) Plot the time series on (i) and arithmetic scale graph,
(ii) a semi-logarithmic graph.

(b) Interpret the results.

Well, the first part of this question will cause you no difficulty at all. As you well know, an arithmetic scale graph is a normal graph, in which we let a given distance (usually a centimetre) represent a given change in the magnitude of the variable on both the horizontal and the vertical axes. We then plot the figures given in the question according to the scale we have used. Thus, in diagram 8.02 on the horizontal axis we could, let two centimetres represent one year, while on the vertical axis it represents £500,000 sales. We have plotted the graphs showing the variation in the sales of the two companies in the normal way.

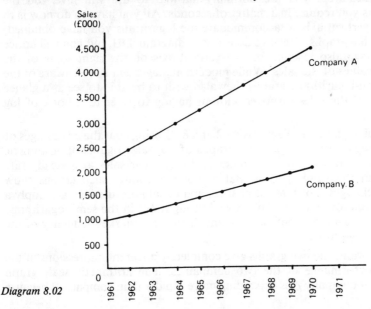

Diagram 8.02

When we now come to consider the semi-logarithmic graphs, we have to be careful. In some examinations you will be issued only with the normal graph paper we have used so far; in others with graph paper already designed for semi-logarithmic use. Let us suppose firstly that you have been issued only with the normal arithmetic scale graph paper. In this case, as you know, we will plot the years along the horizontal axis as normally, but along the vertical axis you will have to plot the logarithms of the sales figure for each year. Our first step them must be to obtain these logarithms. We strongly advise you to do this as part of the answer and not on a piece of scrap paper. It is easy to make a mistake, and if you do examiners are far more inclined to be generous if they can trace easily where the mistake has arisen. So we proceed as follows:

	Company A		Company B	
Year	Sales	Log	Sales	Log
1961	2240	3.3502	980	2.9912
1962	2460	3.3909	1082	3.0342
1963	2680	3.4281	1205	3.0810
1964	2915	3.4646	1289	3.1103
1965	3136	3.4964	1382	3.1405
1966	3362	3.5266	1476	3.1691
1967	3590	3.5551	1580	3.1987
1968	3821	3.5822	1687	3.2271
1969	4049	3.6073	1787	3.2521
1970	4280	3.6314	1891	3.2767

Although this seems to be a cumbersome process, almost any of the small pocket calculators you are normally allowed to use will give you the logarithms you require in a matter of seconds. All you have to do now is to scale the vertical axis to accommodate the logarithms you have obtained, and draw the graphs. We have done this in diagram 8.03, but you will notice that we have also inserted on the vertical axis on the right some of the absolute values of the sales. While modern management is well aware of the use of semi-logarithmic graphs they also wish to be able to see at a glance the value of the sales turnover without having to refer to a book of log tables.

Even a quick glance at diagrams 8.02 and 8.03 brings out the advantages of the semi-logarithmic graph. The graphs of the sales figures on arithmetic scale imply that the two companies increased their sales at a steady rate throughout the whole period. But as you are aware, such graphs show absolute changes, not *rates* of change, and equal absolute increase imply a declining *rate* of increase. This is well brought out in the semi-logarithmic graphs, where a close look at company A shows a marked tendency of the rate of change to fall.

More imporant, the two graphs give completely different impressions of the relative performance of the two companies. The arithmetic scale graph implies that company A is expanding more rapidly than company B because the slope of the graph of the former's sales is steeper. But again the use of absolute figures is deceptive. The semi-logarithmic graphs show that company B's performance is at least as good as that of company A in that it is expanding at the same rate, while on a very close examination we find that the rate of growth over the whole period is marginally better than that of company A, and from 1964 onwards growth certainly seems to be steadier and more sustained than that of company A. Thus, we may come to the conclusion that, although both companies have grown substantially in the period, company B, although smaller, appears to have done marginally the better.

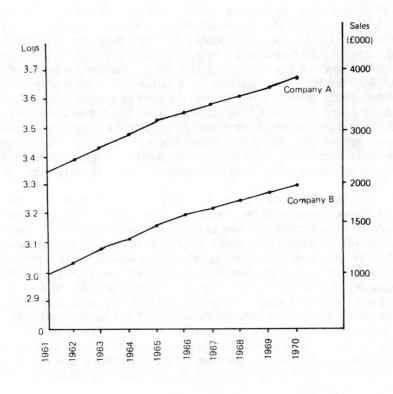

Diagram 8.03

Now this is all very well, but suppose you are issued with semi-logarithmic graph paper already ruled. The great advantage of this paper is that you can insert the absolute figures of sales on the graph without first having to look up the logarithms. Probably, however, few of you have had the opportunity yet of seeing semi-logarithmic graph paper, so let us explain the principles on which it is ruled. One axis is ruled normally, in centimetres and millimetres (after all, it is *semi* logarithmic). But the other axis is quite differently ruled. It is drawn on the basis that equal distances represent equal percentage changes rather than equal absolute changes. Thus, if one centimetre represents a change from 10 to 100 (an increase of 10 times), the next centimetre will represent a change from 100 to 1000 (also an increase of 10 times).

There is more to it than this, however. No-one would expect you to calculate the percentage change for each figure that you have to plot on the graph. We have merely stated the obvious — that one axis of the semi-logarithmic graph paper is scaled logarithmically. Let us take it further. If we take the first centimetre on the vertical scale to represent an increase of 1 unit (from 1 to 2), the next centimetre would represent and increase from 2 to 4. We could have obtained exactly the same effect by plotting logarithms. The log of 1 is 0.0, so you will realise that the logarithmic graph can range upwards only from an absolute value of 1 (not 0 as can the arithmetic scale graph). Now the logarithm of 2 is 0.30103, and the logarithm of 4 is 0.60206. Similarly the logarithm of 8 is 0.90309. Can you see that, in allowing each centimetre to represent the same proportional increase in the absolute figures, we are scaling according to the logarithms of the numbers? In this case one centimetre represents an increase in the logarithm of .30103; and when we add .30103 to a logarithm, we are in fact multiplying the previous number by two. So each successive centimetre represents an increasing change in the absolute figures — 2, 4, 8, 16 and so on.

But the logarithmic scale is not drawn in centimetres as is the arithmetic axis. Suppose once again that we take the first vertical division on the graph paper to be one centimetre long (whether it is or not depends on the graph paper with which you are issued). We can let this centimetre represent any absolute magnitude that we wish. Let us again assume that it represents an increase from 1 to 2. The next main division on our graph paper will also represent an absolute increase of one unit, from 2 to 3, but this division will not be a centimetre deep. We have already said that the second centimetre represents an increase in absolute values from 2 to 4, so you would naturally expect the second division representing an increase from 2 to 3 to be less than a centimetre deep. It will in fact be about .585 centimetres deep only. Thus each successive unit increase on the logarithmic axis will entail a smaller and smaller vertical rise, and the vertical scaling will look something like Diagram 8.04.

In practice, of course, the graph paper you will be given will have each main division subdivided as normal graph paper is into ten subsections to enable you to plot the intermediate figures. If you have not seen such graph paper before, we strongly recommend that you immediately study carefully diagram 8.05 where we have drawn the two graphs we are concerned with on semi-logarithmic paper, and, most important, obtain a stock of such paper of your own and practice drawing semi-logarithmic graphs using the exercises at the end of this chapter.

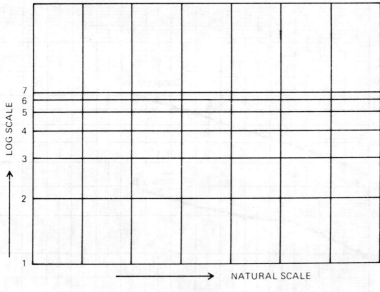

Diagram 8.04

2. The Lorenz Curve

If you look at statistics of income one of the first things that strikes you is the inequality in the distribution of incomes in most countries. Not only is the range of incomes wide, from under £1000 a year to £20,000 a year and more in the United Kingdom, but we also find that a very small percentage of the income recipients at the top of the scale receive a disproportionately large share of total income. Equally, the very large percentage of low income earners receive in total a very small percentage of the total income. You must, at some time in your life, have met such statements as 'the top 5% of income recipients receive over 70% of total income'.

Now any economist will tell you that one purpose of our taxation system, or any taxation system which is progressive, is to reduce the inequality of incomes, and naturally we would all like to know how far the system is succeeding in this objective. Statisticians have derived a diagram, the Lorenz curve, which enables us to show graphically the extent of inequality, not only of incomes, but also of many other things.

In this diagram, considering income distribution, we measure on the horizontal axis the percentage of population, and on the vertical axis the percentage of income, see diagram 8.06.

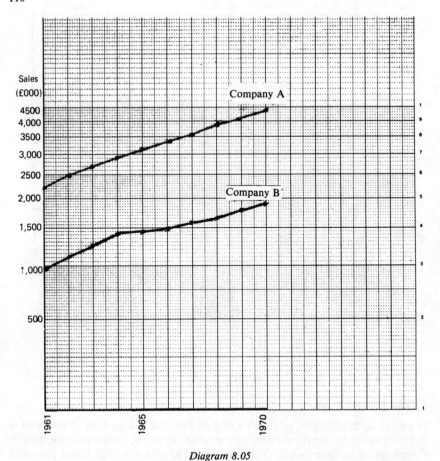

Diagram 8.05

Now, obviously, if incomes are distributed equally, the bottom 10% of income earners will receive 10% of total income, the bottom 20% of income earners will receive 20% of total income and so on. The graph representing such a distribution of income will be the straight line OR passing through the origin and at 45 degrees (provided, that is, that the horizontal and vertical scales are the same), and any divergence from this line will indicate some degree of inequality. The point A for example would be interpreted as 'the bottom x% of income earners receive y% of total income', and, since x is greater that y, would be derived from a situation such as 'the bottom 65% of income earners receive only 32% of total incomes'.

Let us illustrate the use of Lorenz curves by applying them to the following income statistics of the United Kingdom.

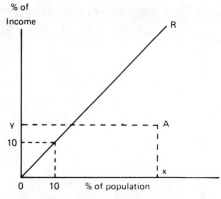

Diagram 8.06

Income Class	No. of Incomes		Pre-tax Income		Post-tax Income	
£	(000)	%	£m	%	£m	%
50–249	5070	18.6	991	4.9	990	5.5
250–499	6570	24.2	2590	12.8	2486	13.8
500–749	6155	22.6	4143	20.5	3844	21.4
750–999	4830	17.8	4580	22.6	4168	23.2
1000–1999	4145	15.2	5849	28.9	5113	28.5
2000–3999	353	1.3	1305	6.4	940	5.2
4000–5999	59	0.2	469	2.3	281	1.6
6000 and over	18	0.1	330	1.6	129	0.8
	27200		20257		17951	

Source: National Income and Expenditure

The first step in the construction of our diagram is to calculate the percentages appropriate to each group and each column. Thus the 5,070,000 individuals in the income class £50 – £249 comprise 18.6% of all income recipients and they received 4.9% of all pre-tax income. We have inserted the relevant percentages in the body of the table, although in an examination you would have to calculate each of them from the original figures given to you.

The next step is to cumulate the percentages you have calculated in this way:

Income Earners (%)		Pre-tax Income (%)	Post-tax Income (%)
18.6	receive	4.9	5.5
42.8		17.7	19.3
65.4		38.2	40.7
83.2		60.8	63.9
98.4		89.7	92.4
99.7		96.1	97.6
99.9		98.4	99.2
100.0		100.0	100.0

It is now easy to draw the Lorenz curves relating to pre-tax and post-tax income:

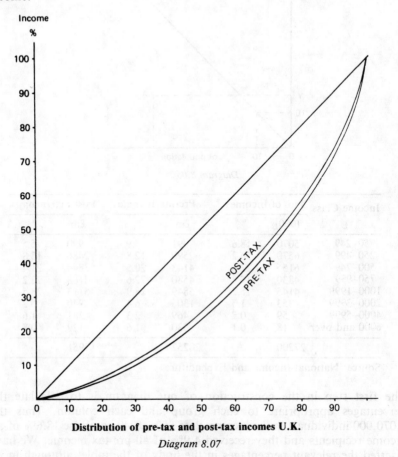

Distribution of pre-tax and post-tax incomes U.K.
Diagram 8.07

Now any divergence from our straight line of equal distribution of income indicates that there is inequality in the distribution of incomes, and the further the Lorenz curve is from this line of reference, the greater is the degree of inequality. It is worth noting that if the Lorenz curve is below the straight line the inequality is in favour of the upper income groups in that a high percentage of low income earners receive a small percentage of income. If, on the other hand, the Lorenz curve is above the straight line, it implies that a given percentage of the bottom income earners receive a higher percentage of total income; the inequality, that is, works in favour of the poor.

Looking at our diagram, it is apparent that at this time there was a great deal of inequality in the income structure of the United Kingdom. Although the taxation system did reduce it, the reduction seems to have been minimal.

We have not, however, asked what was done with the tax revenue. Much of it was returned to the lower income groups in the form of social benefits — family allowances, supplementary benefits, subsidised housing and so on. It may well be that when we allow for this transfer of income, the effects of taxation would appear very different.

3. The Z Chart

A diagram which is often used in industry and commerce, although it seems to be less popular among statisticians, is the Z Chart, so called because the completed diagram takes the form of the letter Z. This is merely a device to enable management to show concisely three different aspects of a time series plotted on the one graph. On the bottom bar of the Z we plot the time series of monthly (or weekly) sales, or output, or whatever variable we are considering. On the diagonal bar of the Z we plot the cumulative total to date, that is, the total sales or output we have achieved since the beginning of the year. Finally as the top bar of the Z we plot the total sales achieved in the last year: the first or January figure is the total sales achieved during the period 1st February last to 31st January this year; the February figure is the total from last March until the end of February this year, and so on.

	Output – ABC Limited			
	Last Year	Current Year	Cumulative Total	Moving Annual Total
January	9	11	11	146
February	8	14	25	152
March	9	12	37	155
April	13	15	52	157
May	14	16	68	159
June	18	19	87	160
July	16	18	105	162
August	15	14	119	161
September	12	14	133	163
October	10	13	146	166
November	9	11	157	168
December	11	15	172	172

We will illustrate by plotting the figures of output for a firm ABC Ltd. The figures are simplified to enable you to follow the calculations more easily. Obviously, although we are plotting figures for the current year only, if we are to obtain a running total of sales over the last twelve months we will need the figures for two years.

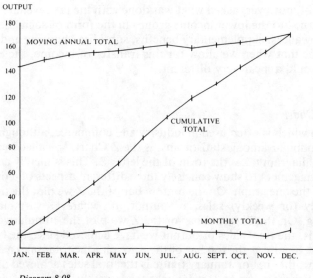

Diagram 8.08

A glance at diagram 8.08 shows what a vast amount of information is given by this simple diagram. The cumulative total tells us performance to date, and it is a simple matter to superimpose on the diagram a line showing planned or expected performance. Thus we can see at a glance whether our plans are being realised or whether we are falling behind. The moving annual total enables us to compare performance this year with that at a comparable time last year. If this chart is rising it means that figures in that month are higher than they were in the same month last year. Thus we have the means of making a direct comparison with last year. Finally the monthly total enables us to keep a direct check on what is happening now; and if we are falling behind our plan we can usually spot a month with low figures which has caused this and so discover why. Is it any wonder, with this wealth of information to be had, that the Z chart is so popular in industry?

4. *The Scatter Diagram*

The final diagram we will introduce you to is not a graph at all. It looks at first glance rather like a series of dots placed haphazardly on a sheet of graph paper. But it is anything but haphazard as we will see. The basic aim underlying the scatter diagram is to try to ascertain if there is a relationship between two factors, such that when one is high the other is high, when one is low the other is low. Or perhaps the relationship is inverse – when one variable is low, the other is high and vice versa. Suppose we were examining the relationship between the level of employment and the level of industrial investment, see Diagrams 8.09a, b and c.

Firstly we will have figures over a considerable period of time giving us the level of industrial investment and the percentage employment rate associated with that level of investment. Let us just take four of those pairs of figures. The first pair tells us that the level of investment is high and associated with it was a high level of employment. This position is indicated by point A in diagram 8.9a. Another pair of figures tells us that when investment was slightly lower employment was considerably lower — point C. Still a third pair of figures tells us that at a time when investment was low employment was also low. This is indicated by point B. If we examine the last figures we find that although investment was the same as before, the level of employment was, in fact, very much higher — point D. If we plot sufficient pairs of figures we may well get a series of crosses such as those on diagram 8.09b which show a pattern. Generally the higher the level of investment the higher the level of employment, and this we have indicated by inserting freehand a dotted line rising upwards from left to right. The same sort of pattern is seen when we examine savings and the rate of interest in Diagram 8.09c. Of course, as you can see from the scatter of the points the relationship is not perfect. We cannot forecast the exact level of employment from the level of investment. It would be fine if we could do this, but at the moment all we are interested in is the general tendency. Before you leave the scatter diagram, experiment for yourself. Draw a scatter diagram showing the relationship between investment and *un*employment. You should get an inverse relationship – the points *falling* from left to right. And finally, try to draw a scatter diagram in which there is no clear relationship shown between two variables.

Some Pitfalls to be Avoided

We must not leave the subject of graphs and diagrams without giving you some advice on what not to do. Unfortunately you will find examples every day of diagrams which illustrate what we are about to say should not happen.

A device beloved of advertisers today is to represent their information in the form of little pictures or ideograms. The sales of a particular brand of beer may be represented by the size of a foaming tankard, the amount of washing-up liquid you get for a penny by the height of liquid in a test-tube. Now, many advertisers using this technique are merely doing their job to the best of their ability; but some advertisements we have seen seem designed to mislead the reader.

Look at diagram 8.10 which represents the sales of 'Whizz' by varying the size of the packet.

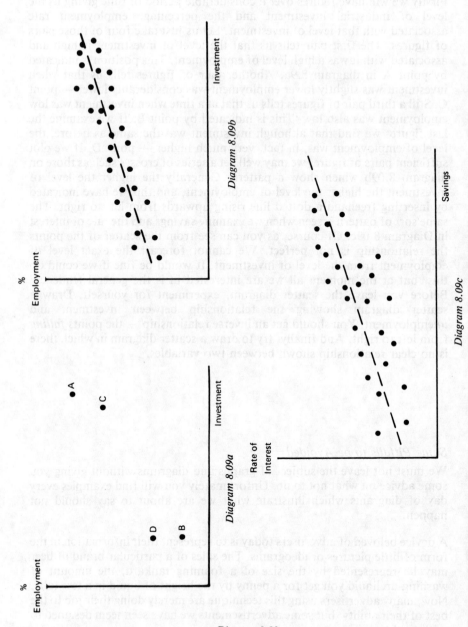

Diagram 8.09

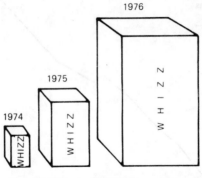

Sales of Whizz are doubling every year

Diagram 8.10

There is nothing wrong with this technique — it can put over the idea of increasing sales forcibly and effectively. But look at the size of the packets. Doubling the dimensions of a packet does not indicate, of course, that sales have doubled. The volume of the second packet is in fact eight times that of the first, and the volume of the last packet is, believe it or not, sixty-four times that of the first. Thus, in spite of the writing saying that sales are doubling every year, the reader is left with a completely false impression of the rate at which they are rising.

Even more unpardonable, in our opinion, is the situation shown in diagram 8.11.

You know that before you draw any graph, you should clearly mark the scales on the axis. Before you indignantly retort that this is an obvious point and will never be forgotten, you must realise that many people do quite deliberately omit the scales, intending to mislead. You must have seen this situation in television commercials. A mysterious line runs across the screen — showing absolutely nothing, but still leaving the impression that sales are skyrocketing and that we are missing the chance of a lifetime by not buying the product. Couple this with the statement that this is the housewives' choice and we are caught. No-one likes to be out of step with one's friends and neighbours.

Equally bad, if not worse, is the invention of units that do not exist. The scales should tell us something; they should use units that are real and that can be understood. But look at Diagram 8.12.

Imagine this graph on your television screen, and the smooth voice of the announcer 'proving' by pointing to the graph that daily brushing with Gritto is bound to make our teeth whiter. But what is a unit of brightness? We are afraid that we do not know, and we very much suspect that the advertising agencies do not know either.

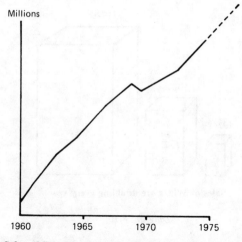

Sales of Bang show it to be the housewives' choice

Diagram 8.11

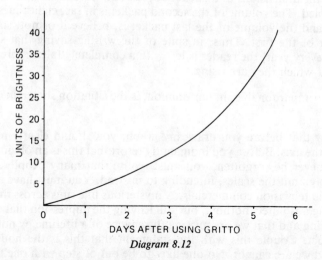

Diagram 8.12

Students often ask us which is the best diagram to use, for a particular purpose. There is no real answer to this and we suggest that you do not waste time looking for one. Any diagram should present the salient features of the data simply and vividly. Equally, if the reader has to spend a great deal of time 'sorting out' what a diagram means, it is a bad diagram.

But do not let it stop here. Certain features of the data should be obvious from the diagram. Now ask questions. Why has the variable behaved in this manner? Why has A behaved differently from B? Is there a relationship between C and D, and, if so, what is this relationship? Only by asking questions such as these will you get full value from your diagrams.

Assignment Exercises

8.1

Transport: Great Britain 1973

	Number of New Registrations of Road Vehicles	Index of Vehicle Distances Travelled (av. 1963 = 100)
January	206,295	153
February	198,531	142
March	223,353	177
April	192,364	180
May	194,446	184
June	186,845	197
July	172,407	214
August	254,438	212
September	159,388	197

Source: Department of the Environment.

(a) The table of transport in Great Britain given above contains two sets of data expressed in different terms. By means of a ratio-scale graph show how these two sets of data may be compared. Use natural scale paper.

(b) Comment on the situation revealed by your graph.

8.2 (i) Describe four diagrammatic methods of presenting numerical data.

(ii) Explain how semi-logarithmic paper differs from arithmetic (or difference) graph paper and give an example of a situation in which semi-logarithmic graph paper would normally be used.

(iii) The following hypothetical data shows the state of weather in London on ten consecutive days.

Day	1	2	3	4	5	6	7	8	9	10
Noon temp. (°C)	19	18	17	20	21	19	18	20	22	19
	C	W	W	S	S	C	W	S	S	C

(C – cloudy, W – wet and S – sunny)

By using appropriate diagrammatic methods, illustrate:

(a) the proportion of days which were cloudy, wet and sunny;

(b) how the noon temperature varied from day to day.

8.3 (a) What advantages has semi-logarithmic graph over a natural scale graph?

(b) Plot the following two series on the same diagram using semi-logarithmic graph paper to show their relative movements and comment on the results.

How could a similar comparison be achieved without using semi-logarithmic paper?

Growth of a Company

	1970	1971	1972	1973	1974
Turnover (£)	70,000	400,000	1,200,000	1,800,000	4,100,000
Cost of materials (£)	30,000	160,000	500,000	750,000	1,400,000

O.N.C.

8.4 (a) For what reasons do we use semi-logarithmic or ratio-scale graph paper rather than the more usual arithmetic scale paper?
(b) Plot the data given below, on semi-logarithmic paper.
(c) Comment briefly on what your graph shows.

Consumer's expenditure in the United Kingdom in £m at current prices from 1960 to 1970

Item/Year	1960	1964	1968	1970
Total consumer expenditure	16,900	21,500	27,200	31,300
Food	4,200	4,900	5,700	6,400
Housing	1,660	2,340	3,290	3,900
Running costs of vehicles	450	780	1,400	1,700

Source: Annual Abstract of Statistics.

8.5 The following table shows the sales of natural gas in the United States, and the natural gas sales of the Metropolitan Gas Corporation, for the period from 1956 to 1968:

Year	United States Natural Gas Sales Cubic Feet (billions)	Metropolitan Gas Corporation Natural Gas Sales Cubic Feet (billions)
1956	7,500	850
1957	8,013	1,030
1958	8,502	1,090
1959	8,750	1,150
1960	9,501	1,250
1961	10,095	1,275
1962	10,700	1,300
1963	11,030	1,350
1964	12,050	1,350
1965	12,800	1,280
1966	13,250	1,302
1967	13,900	1,390
1968	15,500	1,500

(a) Graph this data on a semi-logarithmic scale.
(b) State what the graph indicates about the comparative natural gas sales of the Metropolitan Gas Corporation and the industry as a whole, and explain the advantage of using a semi-logarithmic scale.

A.C.A.

8.6   Using a slide rule or a table of logarithms you are required to construct semi-log graphs and plot on them the following data:

In Great Britain

Year	Total expenditure on highways (£ million)	Total number of cars licensed (thousands)	Total number of goods vehicles licensed (thousands)	Total casualties in road accidents (thousands)
1959	228.0	4,972	1,378	333
1960	238.0	5,532	1,448	348
1961	270.7	5,983	1,503	350
1962	301.2	6,560	1,522	342
1963	342.4	7,380	1,582	356
1964	405.8	8,252	1,633	385
1965	421.2	8,922	1,661	397
1966	457.4	9,522	1,639	392
1967	528.2	10,312	1,692	370
1968	580.9	10,825	1,640	349

Source: Annual Abstract of Statistics.

Interpret the graphs you have drawn and give the advantages, if any, of this type of presentation.   I.C.M.A.

8.7   (a) Plot the following data using arithmetic scale graph paper.

Period	1	2	3	4	5	6	7	8
Data	200	320	640	1180	2080	4050	6480	9030

(b) Describe two methods of drafting a semi-logarithmic curve when no semi-logarithmic graph paper is available. Using one of those methods plot the data and state the advantages and disadvantages of using the semi-log scale over the more usual arithmetic scale.

I.C.M.A.

8.8   *Stoppages of Work due to Industrial Disputes*

	Number of Stoppages beginning in 1973	Aggregate Number of Working Days Lost in these Stoppages (to nearest thousand)
Under 250 days	1200	125
250 and under 500	453	161
500 and under 1000	402	282
1000 and under 5000	592	1250
5000 and under 25,000	180	1844
25,000 and under 50,000	24	859
50,000 days and over	22	2625

Illustrate the above data by means of a Lorenz curve. Why is this form of graph the most suitable for displaying the above information?

8.9 (a) Construct a Lorenz curve in respect of the following data concerning the net output of manufacturing industry X:

Manufacturing Industry X

Average Number of Employees	Number of Firms	Net Output (£ million)
25 and under 100	205	16
100 and under 300	200	60
300 and under 500	35	18
500 and under 750	30	26
750 and under 1,000	20	26
1,000 and under 1,500	10	54

(b) Explain the purpose of presenting information in the form of the Lorenz curve, and comment on your answer to (a) above.

A.C.A.

Chapter Nine

The Averaging of Data

So far, we have concentrated our attentions on organising and presenting data. What we shall now do is attempt to summarise the data at our disposal into a single statistic. This will certainly ease the task of making comparisons, though we must always remember that summarising data is bound to mean that something will be lost.

If you think carefully about it, you will realise that statisticians spend a great deal of their time making comparisons, and the conclusions they reach are often of fundamental importance to every one of us. Comparing income today with income ten years ago is an indicator of how our living standards have changed. Comparing incomes between regions helps the government in its regional policies. Comparing how our prices are changing over time in relation to price changes abroad indicates how competitive our industries are, and comparing our balance of payments over time indicates our ability to pay our way in the world.

Many of the comparisons made present no problems, as we are comparing a single figure value. For example, we can state that I.C.I. earned a certain profit in 1975, and compare this with the profit earned in 1976. Any problems involved in this comparison will be concerned with the calculation of the profit and the rate of inflation, and no-one will dispute that it is legitimate to compare profit in 1975 with profit in 1976 as long as inflation is taken into account. In some cases, however, the things we are trying to compare vary in themselves. Suppose, for example, we attempt to compare incomes on Merseyside with incomes in London: not only is there a variation within the regions, but also a variation within each region. In a previous chapter you saw that we could put raw data into a frequency distribution, and present it in the form of a histogram etc. No doubt we could draw histograms, showing the distribution of incomes in London and Merseyside, and this would enable us to state that incomes earned in London exceeded incomes earned on Merseyside. But is this good enough?

You can probably identify two problems in the statement above. Firstly, do we mean that *all* incomes in London exceed *all* incomes on Merseyside? Obviously not! Many people on Merseyside will earn much more than, say, a porter on the Underground. Secondly, given that incomes earned in London exceed incomes earned on Merseyside, we would wish to know by how much. In other words, we wish to *quantify* the differences in income, and we certainly will not be able to do this by just looking at a frequency distribution or a histogram. What we need is some figure that is *representative* of income in London. We can then obtain a representative figure for Merseyside incomes, compare them, and draw some conclusions as to the size of the difference in incomes.

How are we going to obtain this representative figure? Probably you have guessed already, especially in the current economic climate with our preoccupation with income levels. The TV newscaster does not state that incomes have risen by 10% over the last twelve months: he states that *average* incomes have risen. We are given information on *average* hourly rates of pay, *average* rainfall levels, batting and bowling *averages*, the *average* amount we spend on drink — and so on. The advertising men are very fond of telling us what the *average man* buys, where he goes for his holidays, and what he does with his leisure.

Three conclusions can be drawn from this last paragraph. Firstly, an average is obviously meant to tell us something about the matter under consideration, and unless it is representative of the data, it obviously cannot do this efficiently. Secondly, the word 'average' is one that we meet daily in our conversation, and is a word that is used in a very loose manner. How many times have you heard people use such remarks as 'I *think* that on average I use *about* five gallons of petrol per week'? Here the idea of an average and an *estimate* are shading into each other. We must avoid this at all costs. An average is capable of being calculated from data, and so it is precise. Thirdly, averages are used to describe a wide variety of data, and we must be really sure that we know what an average is. In fact there are many types of average, and we must be sure to select the right one for the right job. If we don't do this, then there is a great danger that the average we quote will not be representative of the data.

The Arithmetic Mean

Most people will tell you to calculate an average something like this: total all the numbers in the group and divide by how many numbers there are in the group. So the average of 5, 7, 9, and 10 is

$$\frac{5+7+9+10}{4} = 7.75$$

In fact, it is easier to demonstrate how to calculate an average than it is to explain how to calculate it. Now mathematicians have developed two useful symbols to overcome this problem. Suppose we put this group of figures we wish to average in a column, and give this column a heading x. The group we considered above would look like this:

$$\begin{array}{c} x \\ 5 \\ 7 \\ 9 \\ 10 \end{array}$$

If we wish to total these figures, the mathematician would state $\Sigma\ x$ – meaning take the sum of the column headed x. (Σ is a Greek capital letter pronounced *sigma*.) Also, we state that there are n figures in the column (in this case $n = 4$). So the expression

$$\frac{\Sigma x}{n}$$

tells us precisely how to calculate the average. We stated earlier that there are many forms of average, and the average we have just calculated is called the *arithmetic mean*. Statisticians use the symbol $\bar{x}$ (pronounce it 'x-bar') to stand for the arithmetic mean, so we can write

$$\bar{x} = \frac{\Sigma x}{n}$$

The arithmetic mean is certainly the most widely used average, both by statisticians and laymen. It will be useful, then, to examine in what sense it is representative. Returning to our example, we have

$$x = 5, 7, 9, 10 \quad \bar{x} = 7.75$$

Notice that the arithmetic mean represents not one single item in the group, so it cannot be representative in the sense that it is typical. It follows, then, that 'representative' means something other than typical. If we subtract the mean from each of the items in the group, we have

$$-2.75, -0.75, +1.25, -2.25$$

We call each of these differences a *deviation* from the arithmetic mean. Notice that the sum of these deviations is zero. Using the sigma notation we have

$$\Sigma(x - \bar{x}) = 0$$

So the arithmetic mean tells us the point about which the values in the group cluster ('mean' in fact means centre, and statisticians call averages measures of central tendency.) This is what we imply when we state that the mean is representative. So we now have a definition of the arithmetic mean – a measure chosen such that the sum of the deviations from it is zero.

For the moment, we shall postpone judging whether this meaning of representative is valid, and concentrate on this important definition of the mean. In fact, the definition enables us in many cases to simplify our calculations of the arithmetic mean. Suppose we guess a value for the arithmetic mean (call this guess x_0). Now if our guess is correct, then the sum of the deviations from x_0 would be zero. If it isn't, then our guess was wrong, and we can adjust our guess to give the true value of the mean. Suppose, for example, we wish to find the arithmetic mean of the group

$$100.1, 100.2, 100.4, 100.8$$

If we guess the mean to be 100, then the deviations are

$$+0.1, +0.2, +0.4, +0.8, \text{sum} = 1.5$$

Clearly, our guess was too low, and we must adjust our guess by

$$\frac{+1.5}{4} = +0.375$$

so the true value of the mean is

$$100 + 0.375 = 100.375$$

(You should check this value by calculating the mean directly.) We can again use the sigma sign to show precisely how to use this method to calculate the arithmetic mean.

$$\bar{x} = x_0 + \frac{\Sigma(x - x_0)}{n}$$

In other words, the arithmetic mean is the assumed mean plus a correction factor.

Now let us examine another factor of the arithmetic mean which will simplify calculations. We can multiply or divide the group of numbers we wish to average, and find the average of this adjusted group. We can then adjust the average we have calculated to the true value. Suppose, for example, we want to find the arithmetic mean of the group

0.0002, 0.0005, 0.0012, 0.0015

Multiplying this group by 10,000 we have

2, 5, 12, 15

which has a mean of 8.5. To obtain a true value for the mean, we now divide by 10,000 giving

$$\bar{x} = 0.00085$$

Of course, in many cases you would not bother to use either of the simplifications mentioned — especially if you have access to a calculating machine. Later, though, we shall meet cases where they speed up our calculations considerably, and also lessen the risk of arithmetic error.

The Arithmetic Mean of a Frequency Distribution

Earlier we recommended that you put raw data into frequency distributions wherever possible, so we must now examine how to find their arithmetic mean. If we consult the General Household Survey, we would learn that 1000 couples married eight years ago would be expected to have the following number of children now.

Number of children	0	1	2	3	4
Number of families	364	362	226	44	4

We require to know the mean number of children per family. First, we will write the data into two columns, one headed x (the number of children —

this is our variable) and the other headed f (frequency). If we multiply the columns together (fx) this will give the total number of children for each family size. So, for example, we see that there are a total of 132 children from 3 children families.

x	f	fx
0	364	0
1	362	362
2	226	452
3	44	132
4	4	16
	1000	962

Adding up the fx column, there are a total of 962 children in the 1000 families, which gives an average of $\frac{962}{1000}$ = 0.962 children per family. Now we know that Σf means total the column headed f (this gives the total number of families) and Σfx means total the column headed fx (which gives the total number of children). So we now know how to find the arithmetic mean of a discrete frequency distribution.

$$\bar{x} = \frac{\Sigma fx}{\Sigma f}$$

Now this is all very well, but much data is in the form of grouped, continuous frequency distributions. Suppose we wished to find the average age of the male labour force. We could obtain the data we require from the Annual Abstract of Statistics.

Age	Number in employment (millions)	Age	Number in employment (millions)
15, but under 20	1.1	45 –	1.5
20 –	1.7	50 –	1.3
25 –	1.5	55 –	1.4
30 –	1.3	60 –	1.1
35 –	1.3	65 but under 70	0.4
40 –	1.4		14.0

Now we have a problem here: look at the age group 20 to 25 years. Within this group there are 1.7 million men, but we have no idea what their *actual* ages are. So we are going to have to make some assumption about their ages. Probably the most sensible assumption to make is that the 1.7 million men in this group have an average age of 22.5 years – the mid-point of the group. If we do this for all groups, then we can use the formula we obtained earlier to calculate the arithmetic mean.

Mid-point x	Frequency f	fx
17.5	1.1	19.25
22.5	1.7	38.25
27.5	1.5	41.25
32.5	1.3	42.25
37.5	1.3	48.75
42.5	1.4	59.50
47.5	1.5	71.25
52.5	1.3	68.25
57.5	1.4	80.50
62.5	1.1	68.75
67.5	0.4	27.00
	14.0	565.00

$$\bar{x} = \frac{\Sigma fx}{\Sigma f} = \frac{565}{14} = 40.36$$

So the average age of working males is 40.36 years.

Now let us see if we can simplify the arithmetic involved in these calculations. Well, the first thing we can do is to multiply all the frequencies by 10: this will remove the decimal fractions and not make any difference to our answer. We can take an assumed mean (in this case we will take 42.5) and calculate the deviations of the items x from 42.5. Notice that all the deviations are divisible by 5, and if we do this we obtain a column that we head d.

x	$(x-42.5)$	d	f	fd
17.5	-25	-5	11	-55
22.5	-20	-4	17	-68
27.5	-15	-3	15	-45
32.5	-10	-2	13	-26
37.5	-5	-1	13	-13
42.5	0	0	14	0
47.5	5	1	15	15
52.5	10	2	13	26
57.5	15	3	14	42
62.5	20	4	11	44
67.5	25	5	4	20
			140	-60

So the average deviation from the mean is $-\frac{60}{140} = -\frac{3}{7}$. But as we divided the deviation by 5, we must multiply the average deviation by 5 to give $-\frac{15}{7}$. In other words, our value for the assumed mean is $\frac{15}{7}$ greater than the true value of the mean. So the true value of the mean is

$$42.5 - \tfrac{15}{7} = 40.36 \text{ years}$$

which agrees with the value we obtained earlier.

Now if we say that d is the deviation from the assumed mean divided by a constant c then we can write a formula for our simplified method like this:

$$\bar{x} = x_0 + \frac{c\Sigma fd}{\Sigma f}$$

Notice that the constant c is normally equal to the class width, though you can choose any value for c which is convenient.

It will be interesting to examine the accuracy of the two means we have calculated. Considering first the mean number of children per family, the distribution gave us the exact number of children in each of the 1000 families. So our calculation of the mean number of children gave us the exact value for the mean: the same result would have been obtained if we had used the frequency distribution or the raw data. However, this is not so with the mean age of the male working population as we did not know the exact age of any of the individuals. In fact, for the members in each class we found it necessary to make an assumption about their ages, so we cannot guarantee the accuracy of the mean. You should realise that although it is much more convenient to consider grouped frequency distributions than raw data, the price we pay for this convenience is the loss of accuracy. In most cases, however, the loss in accuracy is not too serious.

So far we have considered frequency distributions with constant class widths. Now let us look at a case where the class width is uneven.

Distribution of Earnings of Weekly Paid Adults aged 21 and over, April 1971

	Frequency (millions)	Mid-point (x)
£10, but under £16	0.2	14
£16 –	0.4	17
£18 –	0.6	19
£20 –	0.7	21
£22 –	0.9	23
£24 –	1.0	25
£26 –	0.9	27
£28 –	0.9	29
£30 –	1.9	32.5
£35 –	1.4	37.5
£40 –	0.8	42.5
£45 –	0.5	47.5
£50 –	0.4	55
£60 –	0.3	65
£70, but under £100	0.1	85

Source: Annual Abstract of Statistics.

It will certainly be worthwhile multiplying the frequencies by 10, and probably taking an assumed mean (in this case we have chosen £30). However, there is no suitable constant for simplifying the deviations. So we have:

x	$(x-30)$	f	$f(x-30)$
14	−16	2	−32
17	−13	4	−52
19	−11	6	−66
21	−9	7	−63
23	−7	9	−63
25	−5	10	−50
27	−3	9	−27
29	−1	9	−9
32.5	2.5	19	47.5
37.5	7.5	14	105
42.5	12.5	8	100
47.5	17.5	5	87.5
55	25	4	100
65	35	3	105
85	55	1	55
		110	238

$$\bar{x} = 30 + \tfrac{238}{110}$$
$$= £32.16 \text{ per week}$$

So sometimes it is not practical to use the simplified method, and you should always weigh up carefully whether it is worth the labour involved. Better still, buy an electronic calculator (they are very reasonably priced these days) and it will always be worthwhile using the first method.

Limitations on the Use of the Arithmetic Mean

Earlier, we postponed judgement on just how representative was the arithmetic mean. We simply stated that 'representative' certainly doesn't mean typical. The time has now come to put the arithmetic mean through a series of tests, and see how it performs.

If you examine sources of published statistics, you will be surprised how often frequency distributions have 'open-ended' classes. We reproduce below a frequency distribution showing the estimated wealth of individuals in Great Britain.

Wealth	% of Total
Not over £5000	78.38
£5000 − £15,000	16.17
£15,000 − £25,000	2.52
£25,000 − £50,000	1.88
£50,000 − £100,000	0.70
£100,000 − £200,000	0.23
£200,000 −	0.11

Source: Annual Abstract of Statistics.

When we calculated the mean from a frequency distribution, we had to make assumptions about the values of the items in each class. We took the mid-point as the representative value of each class. But what assumptions can we make about the average wealth of individuals in the final class? It would be a brave man indeed who would estimate the upper limit for this group. Likewise, what would be a reasonable lower limit to put on the first class? Unless we have adequate information about the first and last groups, we cannot calculate the mean for this type of distribution. To obtain a mean, we must refer to the raw data (and this just is not available for government published statistics). Clearly, we need a measure that does not depend upon the adequate knowledge of extreme values of the distribution. This problem of open-ended statistics is one you will meet continually in analysis of published statistics — in fact we had quite a job finding distributions that were not open-ended! Before we leave this point, however, it should be stated that if you *must* estimate the missing limit for open-ended distributions, make sure that the frequencies in the open-ended classes are small in relation to other frequencies. Any inaccuracies due to estimates shouldn't, then, be too serious.

The second point about the arithmetic mean is that it often produces results that are not suitable from a communication viewpoint. Earlier, we found that the average number of children per family was 0.962. Most people would consider this to be a ridiculous statement. In situations like this, people expect an integer (i.e. a whole number) to be representative of the number of children per family. If we are going to ensure that all averages are integers, then we must change our idea of an average.

The third point is perhaps the most important of all: the mean is highly sensitive to extreme values. Consider the case of Fred, who at an interview for a job is told that the average income of salesmen in the company is £8000 per year. He accepts the job as he considers the firm to be very progressive with excellent prospects for himself. Although his starting salary is only £2000 a year, his salary will obviously climb very quickly. You could imagine how cheated Fred felt when he found that the sales force consisted of just five men: the sales director on £30,000 a year and four salesmen on £2500 a year! The extreme value (in this case the sales director's salary) has caused the arithmetic mean to be most unrepresentative. To examine this point further, let us return again to the distribution of weekly incomes of males over the age of 21. The cumulative frequency distribution would look like this:

	Cumulative Frequency (millions)
Under £16	0.2
Under £18	0.6
Under £20	1.2
Under £22	1.9
Under £24	2.8
Under £26	3.8
Under £28	4.7

Under £30	5.6
Under £35	7.5
Under £40	8.9
Under £45	9.7
Under £50	10.2
Under £60	10.6
Under £70	10.9
Under £100	11.0

Consulting the ogive for this distribution (Diagram 9.01). We see that 6.45 million people earn less than the average wage of £32.16, which represents nearly 60% of all workers in this category. On this basis, many would claim that the mean is not representative of this data. Had we taken *all* incomes rather than weekly incomes only, then the average would have been much higher (and hence even more unrepresentative). This is the reason that so many people scoff at the average income figures that are quoted from time to time.

Before we search for an alternative to the arithmetic mean, we must say something in its defence. The main redeeming feature of the arithmetic mean is that its calculation involves the use of *all* the data. We will see later that this is not a characteristic of alternatives to the arithmetic mean. In other words, the weaknesses of the arithmetic mean are also its strengths. Because all the data is used, and because the arithmetic mean is capable of precise calculation, most statisticians still prefer it to other measures. In fact the arithmetic mean is capable of, and is the basis for, advanced analysis. Any alternative to the arithmetic mean cannot be used for advanced analysis, so their uses are descriptive rather than analytical.

The Median

Suppose we arranged a group of numbers in ascending order, then the median would be the value of the item in the middle. So the median of the group 3, 4, 7, 9, 11 is 7. Notice that the median is quite unaffected by extremities at either end of the group. Take the case of the sales force considered earlier: the incomes are

£2500, £2500, £2500, £2500, £30,000

and the median income is £2500. The actual size of the director's income makes no difference whatsoever to the median: if the director's salary was doubled or halved then the median would still remain £2500. So the median is quite insensitive to extreme values in the group or distribution, and certainly overcomes the major objections to the arithmetic mean. If you arrange a group of n items in ascending order, then the value of the $(\frac{n+1}{2})$ th item is the value of the median. If we had an even number of items, then there will be *two* middle items. Conventionally, we take the median to be the arithmetic mean of the two middle items. So the median of 4, 9, 13, 17, 30, 32 is $\frac{13+17}{2} = 15$. Now if n is large, the difference between $(\frac{n+1}{2})$ and $\frac{n}{2}$ is negligible — so take $\frac{n}{2}$ in such cases. Finally, then, we can see that half the items in a distribution will be more than the median, and half the items less than it.

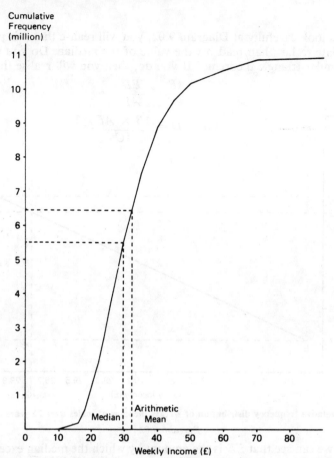

Cumulative frequency distribution of weekly incomes of males over 21 years old 1971

Diagram 9.01

Look again at the cumulative frequency distribution of weekly incomes. There are 11 million workers, so if we wish to find the median income, this will involve finding the income of the 5.5 millionth worker. Reading off the income of this worker from the ogive, we find that the median income is £29.8, compared with a mean income of £32.16. You would probably find that most people would accept the median as being more representative than the arithmetic mean.

It is often quite difficult to read off the value of the median from the ogive, and to overcome this problem it is useful to magnify that part of the ogive containing the median. From the cumulative frequency distribution, we can see that 5.6 million people earn less than £30 per week, and 4.7 million earn less than £28 per week. So the median lies somewhere between £28 and £30 per week. We can plot these points on a graph and join them with a straight line. Reading off from Diagram 9.02, we see that the median income is £29.78.

If you look carefully at Diagram 9.02, you will realise that it is possible to calculate rather than read off the value of the median. Do you remember the similar triangle theorems? If you do, then you will realise that

$$\frac{AB}{AC} = \frac{ED}{AE}$$

so $ED = \dfrac{AB \times AE}{AC}$

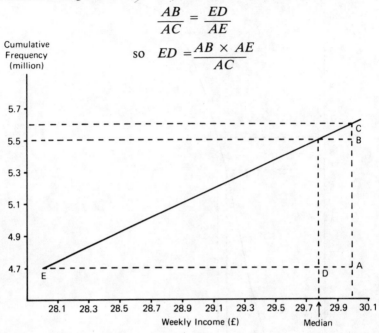

Cumulative frequency distribution of weekly incomes of males over 21 years old 1971

Diagram 9.02

Now we can see that ED is the amount by which the median exceeds 28, so the median is

$$28 + \frac{AB \times AE}{AC}$$

$$28 + \frac{0.8 \times 2}{0.9}$$

$$= 29.78$$

It might be useful to see how we could obtain these figures directly from the cumulative frequency distribution. The relevant parts of this distribution are:

	Cumulative Frequency
Under £28	4.7
Under £30	5.6

so we see that the median is in the group £28—£30, and the median is:

$$28 + \frac{0.8 \times 2}{0.9}$$

Clearly, 28 represents the lower class boundary (LCB) of the median group, and AE represents the class interval of this group. The quantity AC is the frequency of the median group (5.6 – 4.7) and the quantity AB is the median item minus cumulative frequency up to the median group (5.5 – 4.7). So we can calculate the median like this:

$$\text{LCB} + \frac{\text{class interval} \times ([\frac{n+1}{2}] - \text{cum. frequency to median group})}{\text{frequency of median group}}$$

Now let us see if we can calculate the median directly, i.e. without reference to graphs.

Size of Companies Acquired.

Cost	Frequency	Cumulative Frequency
Not more than £100 thousand	279	279
£100 thou. but under £200 thou.	157	436
£200 thou. but under £500 thou.	166	602
£500 thou. but under £1 mill.	117	719
£1 mill. but under £2 mill.	58	777
£2 mill. but under £5 mill.	66	843
£5 mill. but under £10 mill.	30	873
£10 mill. but under £20 mill.	12	885
£20 mill. but under £50 mill.	5	890
Over £50 million	1	891

Source: *Board of Trade Journal*

We wish to find the median cost of companies acquired, i.e. the cost of the $\frac{891+1}{2} = $ 446th company. The median company is in the group £200 thou. – £500 thou., which contains 166 companies. Hence the median is

$$£200,000 + \frac{300,000 \times (446 - 436)}{166}$$

$$= £218,072$$

Notice how difficult it would be to calculate the arithmetic mean of this distribution. It would be extremely difficult to put a lower limit on the first group, or an upper limit on the last group. The median, then, has decided advantages over the arithmetic mean; it can cope with open-ended distributions and is unaffected by extremities at either end of the distribution. Its disadvantage is that it ignores the bulk of the data presented to us, and this disadvantage really is critical! We would like to emphasise again that you should always attempt to use the mean rather than the median, especially if there is not much difference between them. In fact, the difference between them depends on the skewness of the distribution (this is illustrated in Diagram 9.03). The median splits the area under the frequency curve into two halves. If the distribution is symmetrical, then the mean and median will coincide. With a negatively skewed distribution the median exceeds the mean, and with a positively skewed distribution the mean exceeds the median. The more pronounced is the skew, the greater will be the difference between the mean and median. Let us now attempt to summarise this into a simple rule: if you require a representative measure and the distribution is markedly skew, use the median — otherwise use the mean.

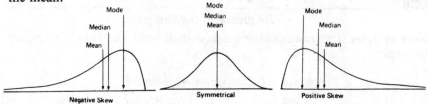

One final problem concerned with the median is that we cannot pool two medians to find an overall median. If we know that a factory employs 100 women at a median wage of £32 per week and 200 men at a median wage of £43 per week, we cannot calculate the median wage for all workers without consulting the raw data. But this is not the case with the arithmetic mean. Suppose the data referred to mean rather than median wages, then the total wages earned by women would be £32 × 100 = £3200, and the total wages earned by men would be £43 × 200 = £8600. So the labour force of 300 earns £11,800 per week, which gives an average wage of $\frac{11,800}{300}$ = £39.33 per week. This pooling of arithmetic means is extremely useful.

The Mode

The mode is the value or attribute that occurs most often, so is an extremely simple concept. Look again at the incomes of the sales force described earlier and you will realise that the modal income is £2500. Straight away, though, we can see a snag with the mode: if all the numbers in the group are different, then we cannot have a modal value.

With a frequency distribution, the mode is the value with the greatest frequency. Let us examine again the distribution of the number of children of 1000 young married couples.

Number of children	0	1	2	3	4
Number of families	364	362	226	44	4

The mean number of children per family we calculated to be 0.962, and the median number of children is one (i.e. the number of children in the 500th family). But the modal number of children is zero, because more families are childless than have any other number of children.

Finding the mode of a grouped frequency distribution will not be quite so easy. Consider again the distribution of weekly paid adults: the class width is not constant so the modal class cannot be obtained by inspection.

Probably the best way to deal with this is to draw the histogram of the distribution. If you examine the distribution carefully, you will see that incomes in excess of £35 are falling off rapidly, so we will concentrate our attention on incomes less than this figure. Consulting diagram 9.04 we can see that the modal class is £24 – £26 per week, and as the adjacent classes have the same frequency, we might be justified in saying the modal income was £25 per week.

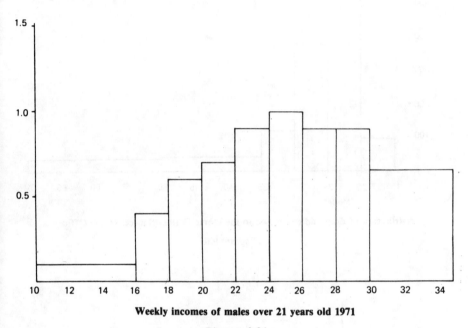

Weekly incomes of males over 21 years old 1971

Diagram 9.04

Suppose the adjacent classes do not have the same frequencies; then we would estimate the mode by splitting the modal class in the ratio of the frequencies of the adjacent classes. This is often done geometrically as illustrated in diagram 9.05.

The mode has the same weaknesses as the median: it ignores the bulk of the data and is not capable of being pooled. We have also seen that in certain groups of numbers, a mode might not be present. Also, in some distributions it is possible to have more than one mode – a distribution like the one in diagram 9.06 is called bimodal. To quote two modes is just clouding the issue. The strength of the mode is that it is extremely easy both as a concept and as a measure of calculation. It is particularly useful to describe attributes – when we state that the average family cleans their teeth with 'Gritto', we mean more families use Gritto than any other toothpaste. You can readily understand why the mode is so popular with **market researchers!**

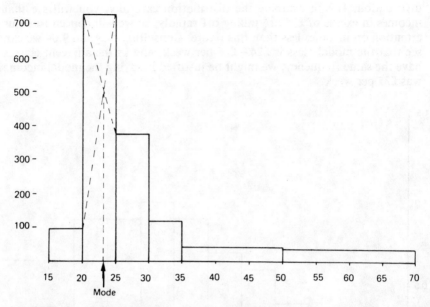

Age distribution of drug addicts known to the Home Office (Source: Home Office)

Diagram 9.05

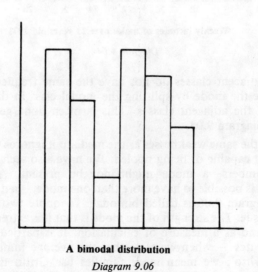

A bimodal distribution

Diagram 9.06

Exercises

9.1 Explain why it is frequently necessary to summarise masses of data by using representative or typical values. Describe two such measures and their methods of calculation. A.C.A.

9.2 If we examine the batting average table for Coalshire we find that Fred Sloggin has an average of 51 runs, and D.E.M. Bones has an average of 39 runs. Would you conclude that Sloggin was the more effective batsman?

Sloggin's average of 51 runs has been obtained in 39 innings. The next innings will be his last of the season. How many runs must he score in the last innings if his final average is to be at least 55 runs?

9.3   *No. of children per 1000 families in which couples were married in 1960-4*

No. of children	0	1	2	3	4	5	6+
Frequency (%)	5.3	5.2	55.4	11.4	18.9	1.5	2.3

Source: General Household Survey.

Calculate the median.

Calculate the mean number of children per family for the above data, given that the average size of a family with 6 or more children is 6.5.

9.4 A company employing 60 people found that the number of sick days taken by its employees last year were as follows:

10	5	12	0	2	35	11	12	4	9
12	17	3	7	8	8	8	10	11	29
44	4	9	3	6	6	7	13	18	4
15	25	5	2	7	20	9	16	10	9
5	2	31	6	0	7	10	9	22	1
3	1	23	9	12	18	6	9	31	0

Group the above figures into intervals of five days. Calculate the mean. Draw the cumulative frequency curve of the distribution and comment on the distribution..

9.5   *Orders Received*

Value of Order (£00)	Number of Orders Received
0 and under 5	20
5 and under 10	51
10 and under 15	139
15 and under 20	116
20 and under 25	31
25 and under 30	14
30 and over	5
	376

(a) From the above table of orders received calculate the mean value of orders received.

(b) Comment briefly on your results for (a).

(c) Suppose the value of the median to be less than the mean. What would this indicate?

9.6 The rateable values of 120 houses were found and the results are shown below.

Rateable Value (£)	Number of Houses
70 – 79.99	3
80 – 89.99	15
90 – 99.99	30
100 – 109.99	36
110 – 119.99	18
120 – 129.99	12
130 – 139.99	6

Calculate the mean of the distribution.

9.7 The table below gives the age distribution of the management of a large company.

Age	Frequency
Under 20	2
20 – 29	12
30 – 39	31
40 – 49	39
50 – 59	26
Over 60	10

Calculate the mean of the distribution. List the assumptions you made in carrying out the calculation and explain why you think you were justified in making them.

9.8 *Age Distribution of the Members of a Golf Club*

Age (in completed years)	No. of Members
10 – 19	185
20 – 29	263
30 – 39	325
40 – 49	442
50 – 59	368
60 – 69	134
70 and over	83

(a) Calculate the arithmetic mean age for the data. (Use mid-value of Group 40—49 as a working origin.)

(b) Define and distinguish between discrete and continuous variables.

9.9 Distinguish between discrete and continuous data. The data below shows the number of local telephone calls made by 75 subscribers during a certain interval of time.

No. of Calls Made	No. of Subscribers
1—10	9
11—15	12
16—20	24
21—25	16
26—40	14
	75

(a) Draw a histogram to illustrate the data.

(b) Calculate the average number of calls made per subscriber.

(c) Calculate the maximum error in your average due to grouping in the data given.

9.10 *Mileages Recorded by 60 Commercial Travellers in the Course of One Week*

515	611	530	557	586	528
533	516	519	560	572	509
520	543	556	532	512	605
559	549	539	609	589	537
524	521	513	541	581	618
544	545	535	568	583	521
555	552	579	581	558	539
562	578	563	598	594	560
595	507	562	532	590	578
526	533	574	531	584	543

(a) From this date tabulate directly a grouped frequency distribution using equal class intervals and starting with '500-under 520'.

(b) Construct the histogram of this distribution.

(c) What is the direction of skew of this distribution?

(d) Calculate the arithmetic mean of the grouped frequency distribution.

(e) Explain why the arithmetic mean of the ungrouped data would be different from the mean obtained in (d).

9.11 The following table shows the age distribution of employees in two factories A and B. Estimate the median age in each factory using an appropriate graph and check the results by calculation.

Age of Employees	Number of Employees	
	A	B
15 – 19	79	5
20 – 24	98	23
25 – 29	128	58
30 – 34	83	104
35 – 39	39	141
40 – 44	19	98
45 – 49	11	43
50 – 54	7	19
55 – 59	3	6

9.12 Explain the difference between a continuous variable and a discrete variable.

The table below gives the earnings (to the nearest £) of 150 employees in a large factory

Income	Frequency	Income	Frequency
£5000 – £5499	1	£2500 – £2999	8
£4500 – £4999	0	£2000 – £2499	34
£4000 – £4499	1	£1500 – £1999	38
£3500 – £3999	2	£1000 – £1499	42
£3000 – £3499	4	£500 – £999	20

(a) What are the class boundaries of the class with frequency 8?

(b) Draw a cumulative frequency curve of the distribution.

(c) What percentage of the group earn less than £2200?

(d) Calculate the median income as accurately as you can.

(e) Under what conditions is the median a better measure of central tendency than the arithmetic mean?

9.13   *Distribution of Personal Incomes before Tax in the U.K. 1969-70*

Income Range (£)	Number of Incomes (00,000's)
below 400	7
400 and under 600	26
600 and under 800	27
800 and under 1000	26
1000 and under 1250	32
1250 and under 1500	30
1500 and under 2000	42
2000 and under 3000	19
3000 and over	8
Total	217

Source: Social Trends, 1971 modified.

(a) Obtain the cumulative frequencies.

(b) By calculation or graphically, obtain the median.

(c) Would you expect the arithmetic mean of the distribution to be greater than, equal to, or less than the median?

Explain you answer. (Do not carry out any further calculations.)

9.14 (a) Define the median, mean, and mode listing the advantages and disadvantages attributable to each.

(b) (i) Calculate the median, mean and mode for the following:

Wage Groups (hourly rate in pence)	Number of Employees
50 and under 60	5
60 and under 70	25
70 and under 80	134
80 and under 90	85
90 and under 100	69
100 and under 110	43
110 and under 120	34

(ii) Illustrate a use for each our the three statistics calculated.

I.C.M.A.

9.15 The marks scored by students in an examination are shown below. (Marking is in whole marks only.)

Marks Scored	10-19	20-29	30-39	40-49	50-59	60-69	70-79	80-89	90-99
No. of Students	4	12	23	37	43	32	19	8	2

(a) Draw a histogram to illustrate the above distribution.

(b) What are the class boundaries of the modal class?

(c) Estimate by calculation the median mark.

(d) If 16⅔% of the pupils are to be given a pass with distinction, what mark will be necessary to achieve this?

(e) What proportion would fail if the pass mark were 48?

Chapter Ten

The Time Series — I

The components of a Time Series

If Plumpton Rovers were drawn against Arsenal in the third round of the F.A. cup, there is no doubt who would be expected to win. Past records show that non-league clubs seldom win against First Division teams. If, however, the non-league club had won such a tie on eighteen out of the last nineteen meetings, then in spite of their relative strength we would expect Arsenal to lose. We tend to base our forecast of the match result on what has happened in the past. This is as true of the accountant as it is of the football fan. In trying to budget for the cost of raw materials or for labour cost in the immediate future, he will look carefully in the first instance at what has happened in the recent past — expecting that the past behaviour of costs will be carried forward into the future.

The statistical series which tells us how data has been behaving in the past is the *Time Series*. It gives us the value of the variable we are considering at various points in time — each year for the last fifteen years; each quarter for the last five years. Yet, when you look at a typical time series, the data fluctuates so much that it seems unlikely that it can help us a great deal. Let us take a series, then, and begin by assessing the factors which cause this fluctuation in the value of the data.

Imports of Raw Material into Ruritania
(1968 = 100)

Year	Quarter			
	1	2	3	4
19-2	114	142	155	136
19-3	116	150	153	140
19-4	128	158	169	159
19-5	137	180	192	172
19-6	145	194		

The figures are given in Index Number form, and they are plotted as a graph in Diagram 10.01.

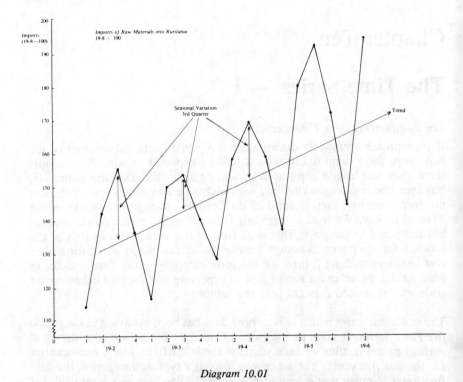

Diagram 10.01

Look carefully at this diagram. Although the figures are fluctuating quarter by quarter two things are immediately apparent. There is a general upward *trend* in the figures as a whole. It is not an exceptional rise, but it is quite marked. Apart from the third quarter of 19-3, the figures in any quarter are higher than those in the same quarter of the previous year.

Secondly, although the figures fluctuate, there is a pattern. The index of imported raw materials is always highest in the third quarter of the year, and always lowest in the first quarter. There is, then, a very marked *seasonal fluctuation* in the figures.

Although it is not apparent in the graph, there is likely to be a third influence on the data. We would not expect import figures to remain unaffected by day to day happenings such as disputes in the docks, exchange rate fluctuations or fuel shortages. It is probably something of this nature that has reduced the figure for the third quarter of 19-3. Thus, our data is likely to be affected by what we can call *residual* or *random influences* which cannot be foreseen or pinpointed without a great deal of outside knowledge — but which may be of great importance.

The series we are looking at is, of course, a short one. We could link the trend perhaps to a general expansion of the economy during the upward phase of the trade cycle. If we were to extend the series and consider a far longer period, the upward expansion might reach a peak and the trend would begin to turn downwards as the economy moves towards depression. Thus, the series we examine might be affected to a great extent by *cyclical influences* resulting in a trend such as that in Diagram 10.02.

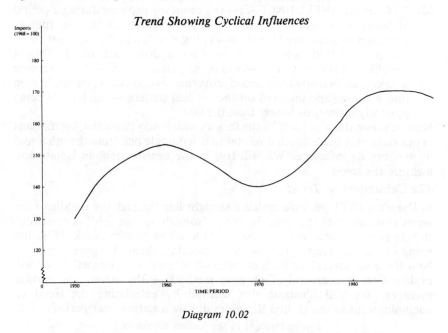

Diagram 10.02

Thus, data such as this which fluctuates quite markedly over time, may be responding to any, or all, of four sets of forces:

(a) The TREND or general way in which the figures are moving. It is important to distinguish between a trend resulting from cyclical influences on the economy and one resulting from, for example, a change in tastes or consumer buying habits. It is likely that there would have been a downward trend in the sales of black and white television sets, even though the economy as a whole was expanding and the sales of all television sets (including colour) were rising.

(b) SEASONAL VARIATIONS. It is common knowledge that the value of many variables depends in part on the time of year we are considering. Every housewife knows that the price of flowers rises as we approach Mothers' Day; or that the price of tomatoes is higher in winter than it is in summer. You can multiply these examples indefinitely from your own experience, and anyone concerned with planning must take account of them.

(c) RESIDUAL INFLUENCES. These are random external events which affect our variables. Sometimes the effect is negligible; at other times it is great; some occurences will increase our figures; others will reduce them. We cannot see what is going to happen in the future, and so we cannot forecast such events. It is, however, reasonable to assume that in the long run they will tend to cancel each other out, and that in our analysis we may initially ignore their impact.

(d) CYCLICAL INFLUENCES. As the economy expands during a period of boom we would expect to find that such data as sales, output or consumer expenditure also show a rising trend; and during a period of slump we would expect them to show a downward trend. Thus a wavelike motion may be observed in the pattern of our data. If we were to take a sufficiently long period, covering several cycles, we might even find a trend superimposed on the cyclical pattern — each cycle being generally higher, or lower, than the last.

Now, it is one thing to identify the factors which may cause the fluctuations in our data: it is quite another matter to disentangle one from the other and to measure its influence. We will try, in the next section, to isolate and measure the trend.

The Calculation of Trend

In Diagram 10.01 we have drawn a straight line through the middle of the series and said that the trend is rising "something like this". For serious statistical work, however, it is no use proceeding by guesswork. If we are going to use the trend figures we must calculate them, not guess them.

Now there are several methods of calculating trend, but the method we will explain first is a general all-purpose method — the method of *moving averages*. We will illustrate this method by calculating the trend of unemployment in the United Kingdom during a sixteen year period.

Unemployment in the United Kingdom

Year	Unemployment %	5 year Total	5 year Moving Average = Trend
1	1.3		
2	1.1		
3	1.2	7.1	1.42
4	1.4	8.0	1.60
5	2.1	8.5	1.70
6	2.2	8.8	1.76
7	1.6	9.4	1.88
8	1.5	9.8	1.96
9	2.0	9.2	1.84
10	2.5	9.0	1.80
11	1.6	9.0	1.80
12	1.4	9.4	1.88
13	1.5	9.3	1.86
14	2.4	10.1	2.02
15	2.4		
16	2.4		

Source: Employment & Productivity Gazette

A moving average is a simple arithmetic mean. We select a group of numbers at the start of the series, in this case the first five, and average them to obtain our first trend figure,

$$(1.3 + 1.1 + 1.2 + 1.4 + 2.1) \div 5 = 7.1 \div 5 = 1.42$$

This trend figure is placed opposite the centre of the group of five numbers, that is, opposite year 3.

To obtain our second trend figure we drop the first number of this initial group (1.3) and include the next number in the series (2.2). So our second group of five numbers is

$$1.1 + 1.2 + 1.4 + 2.1 + 2.2 = 8.0$$

and our second trend figure is

$$8.0 \div 5 = 1.60$$

which is placed opposite year 4. We now drop the figure for year 2 (1.1) and include that for year 7 (1.6), giving us a third trend figure of 1.70. We carry on in this way until all our data is exhausted.

You will notice that, using this method, there is no trend figure for the first two years nor for the last two. Do not forget this if you are asked to plot the original data and the trend on the same graph. Look carefully at Diagram 10.03 where the data have been plotted.

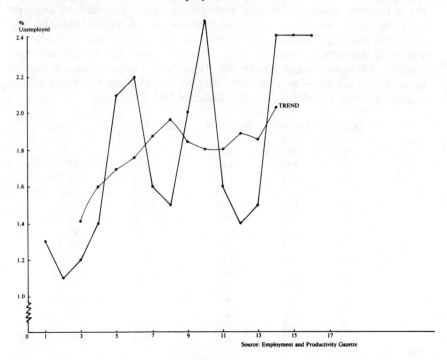

Unemployment in the U.K.

Source: Employment and Productivity Gazette

Diagram 10.03

The trend is probably not as smooth as you thought it would be. True, the large scale fluctuations have been eliminated, and up to year 8 it is fairly smooth; but from this point on there is no clear indication of which way the trend is moving. This was probably a result of changes in government policy as the United Kingdom struggled to counter inflation, prevent large scale unemployment and promote economic expansion.

You are probably wondering why we chose a five year moving average rather than a three year or a seven year one. There is no infallible rule you can follow. The correct time period to use is the one that gives the smoothest trend. A good working rule, however, is to look at the peaks and the troughs in the series. Troughs occur in years 2, 8 and 12; peaks occur in years 6, 10 and 14-16. It looks as if unemployment was subject to a five year cycle, and, in fact, economists confirm that this was so. Hence we chose a five year moving average.

The Trend of a Quarterly Series

A great deal of the data statisticians deal with is given quarterly or monthly, rather than annually. We saw in the previous section that one of the characteristics of such a series is that peaks usually occur in the same quarter of every year, as do the troughs; so the correct moving average to use is the four quarterly moving average. This does not demand any new concepts, but it does raise a practical problem. When we calculate the four quarterly moving average and it is placed opposite the mid point of the group of numbers to which it refers, it falls *between* the second and third figures. Thus we cannot relate our figures of trend to any particular quarter. To see how we overcome this let us calculate the trend of the quarterly series of Imports of Raw Material into Ruritania.

Year	Quarter	Imports	Sum in 4's	Sum of two 4's	Trend
19-2	1	114			
	2	142	547		
	3	155	549	1096	137.0
	4	136	557	1106	138.25
19-3	1	116	555	1112	139.0
	2	150	559	1114	139.25
	3	153	571	1130	141.25
	4	140	579	1150	143.75
19-4	1	128	595	1174	146.75
	2	158	614	1209	151.125
	3	169	623	1237	154.625
	4	159	645	1268	158.5
19-5	1	137	668	1313	164.125
	2	180	681	1349	168.625
	3	192	689	1370	171.25
	4	172	703	1392	174.0
19-6	1	145			
	2	194			

Notice that when we add up in groups of four, the total is placed midway between the relevant quarters. In itself there is little wrong with this, but it prevents any comparison between the level of imports and the trend — a comparison we will have to make if we are to continue the analysis. To eliminate the problem, we use a technique known as *centreing*. At present we have a trend figure of $547 \div 4 = 136.75$ placed opposite quarter 2½, and one of 137.25 placed opposite quarter 3½. If we take the average of these two it would be placed opposite quarter 3, and we have centred the trend so as to relate it to a specific quarter, i.e.

$$\frac{136.75 + 137.25}{2} = 137.0$$

The easiest way to do this in practice is to total successive pairs of four quarterly totals (eg. 547 + 549; 549 + 557) and divide the resultant totals by eight to give the trend. This is done in the last two columns of the table.

Assignment Exercises

Assignment Exercises followed by the letter x do NOT have answers shown at the back of the book.

10.1 Describe the components into which statistical data relating to business and economic events may be analysed by the use of time series analysis.
A.C.A.

10.2 Specify and describe the various movements which may be identified when analysing economic and business statistics over a long time span.
A.C.A.

10.3 In the analysis of a time series:

(a) explain what is meant by a trend line and describe very briefly the method of calculating it.

(b) explain what is meant by residual variations and why they are important;

(c) sketch a graph of a time series which shows a trend and also cyclical and seasonal movements.
A.C.A.

10.4 The table shows the number of parcels carried by Rapid Carriage Ltd. to destinations within the United Kingdom:

Year	Number of Parcels Carried (000's) Quarters			
	1	2	3	4
19-5				800
19-6	690	1580	2250	940
19-7	830	1900	2670	1130
19-8	1040	2160	3070	1360
19-9	1190	2580		

Plot the data and obtain the moving average trend. Plot the trend line on the same graph. What conclusions may be deduced from your graph?

10.5 (a) State four components of a time series.

(b) The following table shows the quarterly sales of a company in thousands of tons for a period of four years:

Sales	Quarter			
	1	2	3	4
Year 1	70	41	52	83
2	78	44	48	85
3	83	54	51	96
4	85	49	54	89

You are required to use the information given above to:
(a) plot on a graph the quarterly sales;
(b) derive and plot the appropriate moving average.

10.6 A traffic census taken at the same time each day during July on a busy road approaching a holiday resort revealed that the number of vehicles passing per hour was as follows (to nearest 10).

	Week				
	1	2	3	4	5
Monday		840	840	830	820
Tuesday		860	860	830	
Wednesday		1190	1200	1220	
Thursday		840	830	840	
Friday		970	1020	1080	
Saturday	1800	1860	1950	2100	
Sunday	1460	1480	1520	1550	

(a) Plot the data on a graph.
(b) Calculate a suitable moving average and plot the trend line. Explain your choice of time period.
(c) If the survey had been carried out in say November, what do you think the figures would have looked like then? (You may illustrate by a rough sketch or by providing possible figures for one or two weeks.)

10.7 Cost of Raw Materials

Year	Cost (£ per ton)
1	3.38
2	4.11
3	4.17
4	3.87
5	3.26
6	3.81
7	3.89
8	4.45
9	4.30
10	3.65
11	5.00

(a) Construct a five year moving average, graphing the trend line against the original data. Comment on your results. If the interval for the moving average had not been suggested, how would you have determined it? What is the purpose of a moving average?

(b) With which characteristic movement of a time series would you mainly associate each of the following?
   (i) a minor fire delaying production for three weeks;
   (ii) an increase in unemployment during the winter months;
   (iii) the increasing demand for small cars;
   (iv) a recession.

10.8 Unemployment as a Percentage of Total Employees

	March	June	September	December
19-5	2.3	2.4	2.6	2.6
19-6	2.9	3.2	3.6	3.7
19-7	3.9	3.5	3.6	3.2
19-8	2.8	2.6	2.4	2.1
19-9	2.4	2.5		

(a) Smooth this time series by means of a centred four-quarterly moving average.
(b) Plot the original and the moving average figures on the same graph.
(c) Discuss whether this moving average has any value for forecasting future percentage unemployment.

Chapter Eleven

The Time Series — II

The Calculation of Seasonal Variation

We have defined seasonal variation as an upswing and downswing in the value of our data. If we are to measure the magnitude of these fluctuations we must have a point of reference from which to measure. After all, when we say a mountain is 3000 feet high, what we really mean is that it is 3000 feet *above sea level*. It seems logical that we should measure the magnitude of the seasonal swing as the deviation from our calculated trend figure. So seasonal variation is not merely an upswing and downswing, — it is a swing around the trend line. Armed with this definition we can now calculate the quarterly variation in imports from our figures. The variation in any particular quarter will be

Original data *MINUS* calculated trend

Year	Quarter	Imports	Trend	Deviation from Trend
19-2	1	114		
	2	142		
	3	155	137.0	+ 18.0
	4	136	138.25	− 2.25
19-3	1	116	139.0	−23.0
	2	150	139.25	+ 10.75
	3	153	141.25	+ 11.75
	4	140	143.75	− 3.75
19-4	1	128	146.75	− 18.75
	2	158	151.125	+ 6.875
	3	169	154.625	+ 14.375
	4	159	158.5	+ 0.5
19-5	1	137	164.125	− 27.125
	2	180	168.625	+ 11.375
	3	192	171.25	+ 20.75
	4	172	174.0	− 2.00
19-6	1	145		
	2	194		

The meaning of the deviations we have calculated in the last column is that in that particular quarter seasonal and other influences have caused imports to vary from trend by the calculated amount. Thus, in the 3rd quarter of 19-2 these influences cause imports to rise 18.0 point above trend; in the 3rd quarter of 19-3, 11.75 points above trend; in the 3rd quarter of 19-4, 14.375 points above and so on. Why should the deviations be so different in the same quarter of the year? The answer lies in the fact that seasonal and *other* influences have been at work. The deviations have been caused by both seasonal and residual influences.

While we cannot separate these two influences, it is reasonable to believe that while seasonal influences will always operate in the same direction, residual influences will sometimes raise the figures and at other times will lower them. If, then, we take a sufficiently long series and take the *average* deviation for any particular quarter, the residual influences will tend to offset each other and we will be left with the purely seasonal variation. This is the rationale behind the calculation of seasonal variation, and the reason why it is often called *average seasonal variation*.

We will now pick up the quarterly deviations from trend and tabulate them in order to calculate seasonal variation.

| | \multicolumn{4}{c}{Quarter} | |
Year	1	2	3	4	
19-2			+18.0	−2.25	
19-3	−23.0	+10.75	+11.75	−3.75	
19-4	−18.75	+ 6.875	+14.375	+0.5	
19-5	−27.125	+11.375	+20.75	−2.0	
Total	−68.875	+29.0	+64.875	−7.5	
Average	−22.958	+ 9.667	+16.219	−1.875	= +1.053
Adjust	− 0.263	− 0.263	− 0.263	−0.263	
	−23.221	+ 9.404	+15.956	−2.138	
S.V.	−23	+9	+16	−2	

There are several things to note about this table. Firstly, you will see that there are four deviation figures for the third and fourth quarters, but only three for the first and second. This is common and depends simply on the lengths of the series we are examining, but we must not forget it when we are calculating the average quarterly deviation. Secondly, the total of the average deviations should be zero, (remember the definition of the arithmetic mean). In fact, the total is +1.053, and since we know it should be zero an adjustment is necessary. We do not know where the difference springs from, so we adjust each quarterly average by the same amount, 0.263 (i.e. 1.053 ÷ 4). Since the total deviations are too many we will have to subtract 0.263 from each quarterly average. Thirdly, when we have made all these adjustments, we are left with average deviations correct to three places of decimals. Now this is silly! In the course of our calculations we have made a number of assumptions. Each one is a logical, reasonable assumption, but we must not pretend to a degree of accuracy we cannot guarantee. Since the original data is given in integers only, it is better to give average seasonal variation in integers also. So we will quote the seasonal variation as −23, +9, +16, −2. Notice that the total still comes to zero.

What do these figures mean? Using the figures for the first quarter as an example, it merely says that economic conditions are such that the index of imports will tend to fall below the trend, and that on average it will be 23 points below. On the other hand, in the second quarter of the year, the economic climate is different, imports tend to rise above trend and on average are 9 points above it.

Series with seasonal variation eliminated

Useful as a knowledge of seasonal variation may be to a firm planning for the future, there is little doubt that the constant fluctuations tend to hide the underlying behaviour of the variable. Comparisons and assessments of performance over time are difficult to make. When a particular salesman returns with a full order book, is it because he has made a superhuman effort, or merely because the seasonal swing is in his favour. The need to make assessments of this nature has resulted more and more in data being produced "seasonally adjusted", or "with seasonal variation eliminated".

Adjusting data to eliminate seasonal variation is a relatively simple matter. If seasonal variation is +23, we are, in effect, saying that our data is 23 points above the trend because of seasonal influences, and we must, therefore, reduce our figures by 23 to eliminate these influences. Similarly, if seasonal variation is −23, we must add 23 to our figures. If we remember that (--) = (+), we may formulate the rule that

Series with seasonal variation eliminated = original data − seasonal variation.

Applying this rule to the data we are analysing:

Year	Quarter	Import Index	Seasonal Variation	Series with seasonal variation eliminated
19-2	1	114	−23	137
	2	142	+9	133
	3	155	+16	139
	4	136	+2	138
19-3	1	116	−23	139
	2	150	+9	141
	3	153	+16	137
	4	140	−2	142
19-4	1	128	−23	151
	2	158	+9	149
	3	169	+16	153
	4	159	−2	161
19-5	1	137	−23	160
	2	180	+9	171
	3	192	+16	176
	4	172	−2	174
19-6	1	145	+9	168
	2	194	+16	185

We now have three different series obtained from the same data, the original figures, the trend and the series with seasonal variation eliminated. When you eliminate the seasonal variation, you might, of course, expect to be left with a series approximating to trend. Do not forget, however, that residual influences will still be found in this series, and the more important the residual influence, the more the series will differ from trend. In Diagram 11.01 we have plotted the three series on the same graph. What conclusions can we draw? Firstly, the trend shows us that there has been a steady and unbroken rise in the volume of imports over the whole period, which shows little signs of ceasing. This may be a result of an expanding industry requiring more and more raw materials, or it may be a result of the domestic consumer buying more and more foreign consumer goods. The graph will not tell us why something happens — but it will lead us to ask the questions.

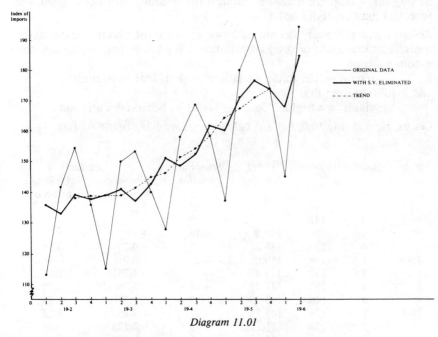

Diagram 11.01

Secondly, the trend and the series with seasonal variation eliminated tend to run closely together. There are differences, but they are small. So we can come to the conclusion that residual influences are fairly unimportant and do not drastically affect the level of imports. This may seem strange, but in fact events such as a dock strike which largely cuts off the flow of goods are very rare.

Nor do such residual influences as there are always work in one direction. Sometimes they raise the level of imports above trend; at other times they lower them; and the pattern is regular. In fact, so regular is the pattern that one can speculate that residuals here have a double effect. At first they slow down the level of imports, but fairly quickly a back-log of orders builds up, and as a result, in the next quarter imports have to rise as the back-log is cleared.

The importance of residuals

You can find out a great deal about residual influences from a graph such as Diagram 11.01, but it does not tell us why residual influences are important.

All planning, whether financial or not, is based on forecasting, and if the forecast cannot be relied on planning is a waste of time. Now, the characteristic of a residual is that it cannot be included in our forecast, so, when its influence is felt, the forecast is upset. In analysing our series of imports we have already discovered that the influence of residuals is fairly small. So, it is probable that though our forecast may be upset, we will not be very far wrong. On the other hand, if the residual influence is great, our forecasts may be badly upset.

Before we come to any decisions, however, we must obtain a quantitative measure of the extent of residual variations. We know that the original data is composed of
　　　　Trend + Seasonal Variations + Residual Variations
and it follows then that,
　　　　Residuals = Original Data − Trend − Seasonal Variations.

Let us, then, bring together the calculations we have made so far:

Year	Quarter	Imports	Trend	Seasonal Variation	Residual Absolute	Residual As % of imports
19-2	1	114		−23		
	2	142		+9		
	3	155	137.0	+16	+2	1.3
	4	136	138.25	−2	−0.25	0.18
19-3	1	116	139.0	−23	0.0	—
	2	150	139.25	+9	+1.75	1.17
	3	153	141.25	+16	−4.25	2.78
	4	140	143.75	−2	−1.75	1.25
19-4	1	128	146.75	−23	+4.25	3.32
	2	158	151.125	+9	−2.125	1.34
	3	169	154.625	+16	−1.625	0.96
	4	159	158.5	−2	+2.5	1.57
19-5	1	137	164.125	−23	−4.125	3.01
	2	180	168.625	+9	+2.375	1.32
	3	192	171.25	+16	+4.75	2.47
	4	172	174.0	−2	0.0	—
19-6	1	145		−23		
	2	194		+9		

Our calculations confirm what we suspected from the graph. The residual influence never affects our figures by more than 4.75 points, and, what is more important, on only two occasions is there more than a 3% influence. Thus, if our forecast is otherwise accurate, we would expect it to be correct to within 3% in spite of the residual influences.

Forecasting from the time series

The key words in the last paragraph are "if our forecast is otherwise accurate". Naturally, in looking to the future, absolute accuracy is difficult to attain, but nevertheless, very goods forecasts can be made.

The basis of the forecasting we will do lies in our knowledge of the behaviour of trend. The trend will not suddenly change direction. As you can see, in Diagram 11.02 the trend is beginning to rise rather more slowly. Over the next few quarters the rate of rise may continue to fall and eventually the trend will reach a maximum and perhaps begin to fall. Or it may slowly change direction upwards with an accelerated rate of rise. The point is that it is not going to change *suddenly*. So, we can extend (or project) the trend forward over the next two or three quarters, confident that our projection will be fairly accurate. The further we project the trend, of course, the greater is the possibility of inaccuracy, and great care is needed in making the projection.

In diagram 11.02 the trend is drawn and a suggested projection made, extending over the next two quarters. We make no claim, of course, that this projection is perfect, and you may make a far more accurate one yourself. Our new trend values are:

19-6 Quarter 1   176.5
            2   178.75

To understand just how we can use these figures we will go back to the calculation of trend and reproduce the last part of the calculations.

Year	Quarter	Imports Col. 1	Sum in 4's Col. 2	Sum of two 4's Col. 3	Trend Col. 4
19-5	3	192	681	1370	171.25
	4	172	689	1392	174.0
19-6	1	145	703	($8x = 1412$)	($x = 176.5$)
	2	194	($8x - 703 = 709$)	$8y =$	($y = 178.75$)
	3	($a = 198$)			
	4	$b =$			

We will call the forecast values for 19-6, quarter 3, *a*, and that for quarter 4, *b*. Suppose the trend value we read from the graph is x, – this is placed in column 4. But we obtain x by dividing column 3 by 8, so in columns 3 we must place 8x. In its turn this 8x is the sum of 703 plus an unknown number, which will have therefore a value 8x – 703, which is the next figure in column 2. Column 2 is the sum of four import indices and $8x - 703 = 145 + 172 + 194 + a$. It follows then that

$$a = 8x - 703 - 172 - 145 - 194 = 8x - 1214$$

Since $x = 176.5$

$$a = (8 \times 176.5) - 1214 = 1412 - 1214 = 198$$

This seems to be a reasonable forecast since the third quarter is always slightly higher than the second quarter (which was 194). The values we have calculated have been placed in the appropriate columns in the table. Now, using these figures, and given that the value of the trend y is 178.75, calculate for yourself the value of the import index for the fourth quarter of 19-6.[1] The answer is given in the footnote, and you will probably agree that the two figures we have forecast are entirely consistent with the pattern of the series.

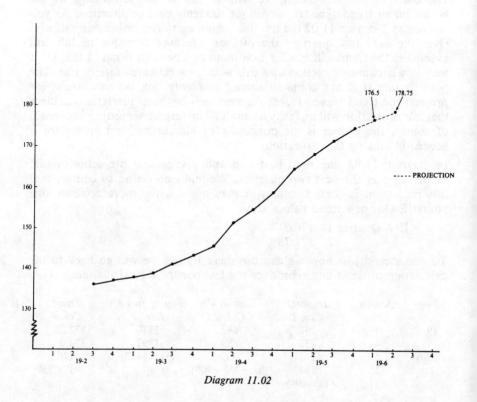

Diagram 11.02

Forecasting by projecting the trend of an annual series is exactly the same as for a quarterly series. Suppose we are taking a five year moving average, and the last few rows of our calculations are:

Year	Output	5-year total	Trend
19-2	15	70	14.0
19-3	18	74	14.8
19-4	13	76	15.2
19-5	14	?	?
19-6	16		
19-7	?		

1. Answer: $y = 178.75$ $b = 8y - 709 - 145 - 194 - 198 = 8y - 1246 = 1428 - 1246 = 182$.

The first step in forecasting a value for 19-7 would be to project the trend forward. We could then read a value for 19-5. Suppose this value of the projected trend were 15.6. This figure is the average of a five year total in the previous column, so that total must be 5 × 15.6 = 78. This in itself is the sum of 18 + 13 + 14 + 16 + (output in 19-7). Thus output in 19-7 would be forecast as

$$78 - 18 - 13 - 14 - 16 = 17.$$

The calculation of seasonal variation in this example is based on the assumption that what we call the Additive Model is the correct one to use. We assume, that is, that

Data = Trend + Seasonal Variation + Residuals

or, as it is more often put,

$$Y_t = T_t + S_t + R_t$$

Now, in making this assumption we arrive at a figure for seasonal variation which is taken to be applicable to that particular quarter in any and every year. There is little harm in this if the magnitude of the swing around the trend is fairly constant. But suppose that over time the magnitude of the swings is getting larger — we have a problem. We might find that in measuring the deviations from trend for the first quarter they read +4, +8, +14, +23. The average seasonal variation is +12.25. If, however, the deviations continue to be of the order of 23 or more, it is nonsense to use a figure of +12.25 in order to forecast data in the future. You must use the additive method only when the deviations from trend which you obtain are fairly stable.

In their calculations of seasonal variations many firms find that the trend of their data is a linear one. Numbering each quarter in order, 1, 2, 3, 4, 5, 6 and so on, they can then express the equation of the trend line in the form y = a + bx. They may find, for example, that the trend of their output data is given by Trend = 6.0 + 2.5t, where t is the number given to the appropriate quarter. Let us study an example using this principle.

Alpha Products Ltd. have produced figures for their export sales over the last four years as follows:

Sales (£ million)
Quarter

1	2	3	4
8	17	10	18
19	24	17	29
23	38	31	33
34	45	42	48

The firm knows that the trend line is given by the equation

$$\text{Trend} = 6.0 + 2.5t$$

Calculate the average seasonal variation and use your figures to forecast sales for each quarter of next year.

If you remember that t is the number given to the particular quarter, the first quarter of the first year being numbered 1, the trend is easy to calculate for that quarter as:

$$6.0 + 2.5 \times 1 = 8.5.$$

In the same way the trend value for the second quarter of that year is

$$6.0 + 2.5 \times 2 = 11.0$$

Our calculations now proceed as follows:

Quarter	Sales	Trend	Deviation from trend
1	8	8.5	−0.5
2	17	11.0	6.0
3	10	13.5	−3.5
4	18	16.0	2.0
5	19	18.5	0.5
6	24	21.0	3.0
7	17	23.5	−6.5
8	29	26.0	3.0
9	23	28.5	−5.5
10	38	31.0	7.0
11	31	33.5	−2.5
12	33	36.0	−3.0
13	34	38.5	−4.5
14	45	41.0	4.0
15	42	43.5	−1.5
16	48	46.0	2.0

Seasonal Variation can now be calculated.

	Quarter				
	1	2	3	4	
	−0.5	6.0	−3.5	2.0	
	+0.5	3.0	−6.5	3.0	
	−5.5	7.0	−2.5	−3.0	
	−4.5	4.0	−1.5	2.0	
Total	−10	20	−14	4	
Average	−2.5	5	−3.5	1	= 0

In using these figures to forecast for the next year we remember that the figure for sales is

$$\text{Trend} + \text{Seasonal Variation}.$$

For the first Quarter of next year our forecast would be
$$6.0 + 17 \times 2.5 - 2.5 = 46.0$$
For the second quarter
$$6.0 + 18 \times 2.5 + 5.0 = 56.0$$
For the third quarter
$$6.0 + 19 \times 2.5 - 3.5 = 50.0$$
And for the fourth quarter
$$6.0 + 20 \times 2.5 + 1.0 = 57.0$$
Alpha Products then would predict:

	Seasonal Variation	Next Year's Sales
Quarter 1	−2.5	46.0
Quarter 2	+5.0	56.0
Quarter 3	−3.5	50.0
Quarter 4	+1.0	57.0

Additive or Multiplicative Model?

In our analysis of time series, we have assumed that

$$\text{Actual data} = \text{Trend} + \text{Seasonal Variation} (+ \text{Residuals})$$

This is the so-called *additive model*.: it assumes that there is a constant absolute difference between the actual data and the trend, and we call this difference the seasonal variation. Now most statisticians think this is not a reasonable assumption: it does not seem reasonable to adjust every (say) first quarter by adding or subtracting a constant amount. It would be much better to adjust every corresponding quarter by a constant percentage. This gives the more satisfactory *multiplicative model*.

$$\text{Actual data} = \text{Trend} \times \text{seasonal variation} (+ \text{Residuals}).$$

Instead, then, of taking seasonal variation to be the average of the (actual − trend) values, we will take it to be the average of the (actual/trend) values. Let us rework our example of the import index, this time using the multiplicative model. The trend is calculated in exactly the same manner as for the additive model − by using a four quarter, centred, moving average.

Year	Quarter	Import Index	Trend	Actual/Trend
19-2	1	114		
	2	142		
	3	155	137.000	1.1313
	4	136	138.250	0.9837
19-3	1	116	139.000	0.8345
	2	150	139.250	1.0771
	3	153	141.250	1.0831
	4	140	143.750	0.9739
19-4	1	128	146.750	0.8722
	2	158	151.125	1.0454
	3	169	154.625	1.0929
	4	159	158.500	1.0031
19-5	1	137	164.125	0.8347
	2	180	168.625	1.0674
	3	192	171.250	1.1211
	4	172	174.000	0.9885
19-6	1	145		
	2	194		

Year	Quarter 1	2	3	4
19-2			1.1313	0.9837
19-3	0.8345	1.0771	1.0831	0.9739
19-4	0.8722	1.0454	1.0929	1.0031
19-5	0.8347	1.0674	1.1211	0.9885
Total	2.5414	3.1899	4.4284	3.9492
Average	0.8471	1.0633	1.1071	0.9873

We would predict the total of the average ratios to be 4, whereas it is 4.0048. So we must adjust the averages by multiplying each average ratio by $^4/_{4.0048}$ to give seasonal variation ratios

Quarter 1	2	3	4
0.846	1.062	1.106	0.986

If we wish to find a deseasonalised series, we would divide the actual data by the seasonal variation.

Year	Quarter	Index	Seasonal ratio	Deseasonalised Index
19-2	1	114	0.846	135
	2	142	1.062	134
	3	155	1.106	140
	4	136	0.986	138
19-3	1	116	0.846	137
	2	150	1.062	141
	3	153	1.106	138
	4	140	0.986	142

19-4	1	128	0.846	151
	2	158	1.062	149
	3	169	1.106	153
	4	159	0.986	161
19-5	1	137	0.846	162
	2	180	1.062	169
	3	192	1.106	174
	4	172	0.986	174
19-6	1	145	0.846	171
	2	194	1.062	183

The Adaptive Forecasting

The use of trend to provide a forecast is an example of what a statistician would call *adaptive forecasting*. Think for a moment about what we have done in the previous sections. We have projected the trend forward for one or two time periods, taking into account all the information which we currently have available. We then used our projected trend to forecast actual data values. Time, however, does not stand still, and very shortly we will have the accurate value of the data for the next time period. Now possession of this data gives us two valuable pieces of information. It enables us to check the accuracy of the forecast we made; and it enables us to calculate the value of the next trend figure. If this new calculated value is very close to the value we projected our forecast will probably be fairly accurate, and there will be little reason for us to change our opinion of the way in which trend is going to behave. But suppose our new trend figure differs markedly from the projected figure. We would be well advised to reconsider our opinion of the future behaviour of trend and to make some adjustments to the projection. In this adjustment to the projected trend we will naturally have to recalculate the forecasts we have made. This, then, is adaptive forecasting. As each new item of information becomes available, we adapt our forecast in the light of the new data we now have. Thus forecasting is constantly revised or adapted as we obtain more data on which it can be based.

As you know, one of the first problems we had to consider when we calculated the trend was that of deciding how many time periods to use. The choice is a critical one for the forecaster. Suppose we decided to use a three year moving average. As you can see from the table, the value of the trend fluctuates and adjusts very quickly to fluctuations in the data. Thus our forecasts based on the trend will also tend to fluctuate as data fluctuates in value. Our forecast will be highly *sensitive* to any variation in the data which may occur. Is this such a bad thing? Well, we want our forecast to take account of variations in data, yet, at the same time we must remember that data may vary for purely random reasons rather than as part of an underlying trend. If our forecast is too sensitive it will be reflecting these random influences rather than trend.

Data	3 Period total	trend	11 period total	trend
60				
52	163	54.33		
51	152	50.67		
49	144	48.00		
44	120	40.00		
27	96	32.00	498	45.27
25	93	31.00	502	45.64
41	108	36.00	499	45.36
42	127	42.33	498	45.27
44	149	49.67	495	45.00
63	171	57.00	506	46.00
64	176	58.67	535	48.64
49	163	54.33	554	50.36
50	145	48.33	569	51.73
46	151	50.33	578	52.55
55	157	52.33	590	53.64
56	155	51.67		
44	156	52.00		
56	151	50.33		
51	163	54.33		
56				

Suppose now we go to the other extreme and consider the calculation of trend using an eleven year moving average. There are still fluctuations in the value of the trend, but you can see from the table that fluctuations in the original data do not affect the value of this trend to anything like the same extent as they affected the three year moving average. There is little sensitivity in the trend but it is far more *stable* than the previous one. This difference is well brought out in the graph of the two trends drawn in Diagram 11.03.

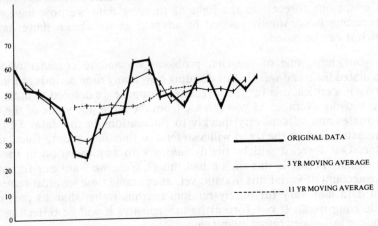

Diagram 11.03.

This, then, is the problem facing the forecaster. He must not provide a forecast which is so sensitive that it reflects every random variation; yet the forecast must not have such a high degree of stability that genuine changes in the underlying trend are not reflected in the forecast. This question of striking a balance between sensitivity and stability arises in any adaptive forecasting technique. If we are using moving averages, as we have seen, it is a question of choosing the right time period. Too short a period and we have extreme sensitivity but no stability; too long a period and we get a great deal of stability with no sensitivity. The balance the forecaster ultimately adopts will depend on the relative importance he himself attaches to sensitivity and stability.

Exponential Smoothing

In spite of its high sounding title *exponential smoothing* is merely an adaptive forecasting technique of great simplicity. Let us suppose that we had made a forecast that our December sales would be £35,000 but that in fact they were £38,000. All that exponential smoothing demands is that our forecast for the January sales be adjusted in the light of the difference of £3,000 between the actual and the forecast figures for the previous month. But what adjustment should we make? It seem logical that it should be some fraction of the difference — a half, or a quarter, or seven eighths and so on. Suppose we represent the chosen fraction by the Greek letter α (alpha), we can say:

New Forecast = old forecast + α (Actual Data − Old Forecast)

where α is a figure lying between 0 and 1.

Let us see how this would work in practice taking a value for α of 0.5.

When initially we begin to forecast our weekly sales, the only information we have is that this week's sales (week 1) were £3,500. The best forecast we can realistically make without any further information is that next week's sales will also be £3,500. In the event we find at the end of week 2 that our sales were in fact £3700. We will now adjust our forecast for week 3 in the light of this additional information. Our adjustment factor, α, is 0.5, so our forecast for week 3 is produced as follows:

New Forecast = Old Forecast + α (Actual Data − Old Forecast)
= 3500 + 0.5 (3700 − 3500) = 3600

So we will adjust our forecast for week 3 by an amount equal to half the error we made in week 2, that is to £3600. It is easy to see that this forecast takes into account all the information in our possession, both for week 1 and week 2.

Suppose now that actual sales in week 3 were £4100 or £500 above the forecast figure. For week 4 we would adjust our forecast further by an amount equal to this difference multiplied by α, that is, by £500 × 0.5. Thus our forecast for week 4 would be:

$$£3600 + 0.5 \times £500 = £3850$$

In tabular form the information would be presented as follows:

Week	Forecast £	Actual Sales £	Difference	Adjustment $\alpha = 0.5$	New Forecast
1		3500			3500
2	3500	3700	+ 200	+ 100	3600
3	3600	4100	+ 500	+ 250	3850
4	3850				

It is apparent that the new forecast we make will depend critically on the value we give to α. Let us consider an extreme case when the value we give to α is 0.0. Is it not obvious that once we have made our initial forecast this forecast will never be adjusted no matter how large the difference between the actual and the forecast figures?

Thus, having forecast sales of £3,500 in week 2 and achieved sales of £3,700, our forecast for week 3 will be:

$$£3,500 + 0.0 (£3,700 - £3,500) = £3,500 + 0 = £3,500.$$

We will never adjust our forecast because an alpha value of 0.0 means that we are taking no account whatsoever of any new information we may receive. We have a forecast which is absolutely stable, but which is completely insensitive to fluctuations in the data.

There is no golden rule which will tell you precisely what value to take for alpha, but in the tables below we have exponentially smoothed a series using a number of different alpha values, and you can see for yourself how sensitivity is gained at the expense of stability.

Actual Data	Forecast	Difference	Adjustment	New Forecast
When $\alpha = 0.1$				
103	72.5	+ 30.5	+ 3.05	75.55
112	75.55	+ 36.45	+ 3.645	79.195
110	79.195	+ 30.805	+ 3.0805	82.2755
135	82.2755	+ 52.7245	+ 5.2724	87.5479
143	87.5479	+ 55.4521	+ 5.5452	93.0931
152	93.0931	+ 58.9069	+ 5.8907	98.9838
150	98.9838	+ 51.0162	+ 5.1016	104.0854
175	104.0854	+ 70.9146	+ 7.0915	111.1769
183	111.1769	+ 71.8231	+ 7.1823	118.3952
192	118.3592	+ 73.6408	+ 7.3641	125.7593

Now as you can see in this table, an alpha value of 0.1 is not very satisfactory. Not only is the forecast consistently less than the actual data, the deficit is getting larger with each successive period. Beginning with a difference of only 30.5, we end up only nine periods later with a deficit of over 73. The trouble is that an alpha value as low as 0.1 has insufficient sensitivity. The forecast will constantly fall further and further below the actual data so long as this trend continues. Diagram 11.04, where both the original data and the forecast are plotted, shows this forcibly.

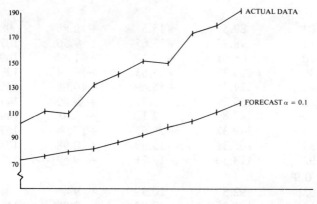

Diagram 11.04.

If an alpha value of 0.1 results in a forecast pattern which is not only bad but which may be misleading, can we improve the situation by increasing the value of alpha and so increasing sensitivity? Remember, though, that in judging the tables which follow that we can only increase sensitivity at the cost of stability.

Data	Forecast	Difference	Adjustment	New Forecast
α = 0.3				
103	77.5	+25.5	+ 7.65	85.15
112	85.15	+26.85	+ 8.06	93.21
110	93.21	+16.79	+ 5.03	98.24
135	98.24	+36.76	+11.03	109.27
143	109.27	+33.73	+10.12	119.39
152	119.39	+32.61	+ 9.78	129.17
150	129.17	+20.83	+ 6.25	135.42
175	135.42	+39.58	+11.87	147.29
183	147.29	+35.71	+10.71	158.01
192	158.01	+33.99	+10.20	168.20
α = 0.5				
103	82.5	+20.5	+10.25	92.75
112	92.75	+19.25	+ 9.63	102.38
110	102.38	+ 7.63	+ 3.81	106.19
135	106.19	+28.81	+14.41	120.59
143	120.59	+22.41	+11.20	131.80
152	131.80	+20.20	+10.10	141.90
150	141.90	+ 8.10	+ 4.05	145.95
175	145.95	+29.05	+14.53	160.47
183	160.47	+22.53	+11.26	171.74
192	171.74	+20.26	+10.13	181.87

$\alpha = 0.7$

103	87.5	+15.5	+10.85	98.35
112	98.35	+13.65	+ 9.56	107.91
110	107.91	+ 2.10	+ 1.47	109.37
135	109.37	+25.63	+17.94	127.31
143	127.31	+15.69	+10.98	138.29
152	138.29	+13.71	+ 9.59	147.89
150	147.89	+ 2.11	+ 1.48	149.37
175	149.37	+25.63	+17.94	167.31
183	167.31	+15.69	+10.98	178.30
192	178.30	+13.71	+ 9.59	187.89

$\alpha = 0.9$

103	92.5	+10.5	+ 9.45	101.95
112	101.95	+10.05	+ 9.05	111.00
110	111.00	− 1.00	− 0.90	110.10
135	110.1	+24.90	+22.41	132.51
143	132.51	+10.49	+ 9.44	141.95
152	141.95	+10.05	+ 9.05	151.00
150	151.00	− 1.00	− 0.90	150.10
175	150.10	+24.90	+22.41	172.51
183	172.51	+10.49	+ 9.44	181.95
192	181.95	+10.05	+ 9.04	191.00

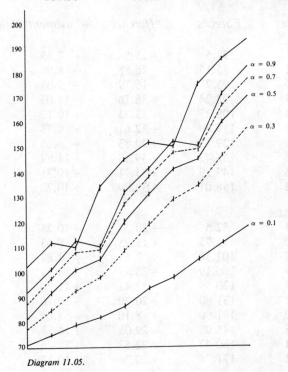

Diagram 11.05.

Now, as you can see from the graphs, if we take an alpha value of 0.9, the forecast we make reflects almost exactly what has been happening to the data in the previous year. Our forecast is highly sensitive, but in fact it reflects every minor variation in the actual data whether that fluctuation is purely random or not. At the other extreme an alpha value of 0.1 does not reflect any of the fluctuations in the original data. So true is this that an alpha value as low as this does not seem to give us a forecast reflecting even the trend. Our forecast figures are falling further and further below the actual data. We have a great deal of stability but no sensitivity. It would seem, from the graphs that we have drawn, that an alpha value of 0.3 would be the best to use. Our forecast will reflect adequately the generally rising trend of the figures, but is does not seem to be influenced by short term random fluctuations in the data. This conclusion, in fact, seems to reflect practice. The normal value of alpha used in this type of forecasting is very very rarely greater than 0.5 while a value of 0.3 is quite usual.

One final point is important. You will doubtless be worried by the fact that the forecasts we have made are *all* well below the actual data in every time period. This has enabled us to separate the graphs we have drawn for the sake of clarity. If you look carefully you will see that the low forecast springs in every case from the fact that our first forecast figure has been assumed to be well below the actual data. So long as there is an upward trend no value of alpha will ever enable us to catch up. In practice it is crucial, then, that if your forecast is to have any credibility the initial forecast must be compatible with the actual data. If it is not, every other forecast you make will reflect the same inaccuracy.

We must not end this chapter leaving you with the impression that forecasting is merely a simple arithmetic calculation. It is an art. Always look at your forecast and ask if it seems logical; if it fits the pattern of the series. If you feel it does not, obey your instinct and look at the trend again. The best forecasters are often those who get a gut feeling as to how the trend will behave a year or two ahead. Above all never place too much reliance on figures forecast for more than a short period ahead. You should realise that there are so many factors affecting the pattern of the time series, that it is unlikely that they will all remain constant or behave normally for any longer than the immediate future. Fortunately it is only rarely that you will be interested in many years ahead. In preparing budgets, in forecasting sales, in assessing how costs will behave, the most important period is – the very next year.

Exercises to Chapter 11
(Use additive model unless stated otherwise)

11.1 The number of prescriptions dispensed by chemists under the National Health Service in England and Wales during a five year period is shown in the table below.

N.H.S. Prescriptions (millions)

Year	Quarters			
	1	2.	3	4
1			60	71
2	69	67	62	69
3	73	66	62	68
4	72	66	65	67
5	75			

(a) Plot the data on a graph.

(b) Calculate and plot on the same graph a suitable moving average.

(c) Explain what additional calculations would be necessary to obtain the average seasonal variation. (Do *not* make these calculations).

11.2

Percentage Turnover of Labour Force

Year	Quarter			
	1	2	3	4
19-0	11	23	16	8
19-1	14	29	16	9
19-2	18	34	17	9
19-3	19	42	23	12

(a) Smooth this time series by means of a centred four-quarterly moving average.

(b) Calculate the average seasonal variations (use the multiplicative model).

(c) Explain how your calculations could be used as a basis for forecasting percentage turnover figures for the four quarters of 19-4.

11.3

Quarterly Sales (£000)

Year	Quarter			
	1	2	3	4
19-0	19	30	63	8
19-1	20	32	67	16
19-2	23	35	74	14

Calculate the average seasonal variation for each quarter of the year and, using the second quarter as an example, state what is meant by your calculated figure.

11.4

Number of Houses completed in England and Wales

Year	Quarter			
	1	2	3	4
19-0	32458	32881	35049	39462
19-1	33310	31953	33628	35983
19-2	30808	27935	27526	30946
19-3	28674			

(a) Rewrite the number of houses completed to the nearest hundred.

(b) Using the rounded data:

   (i) calculate, by means of four quarterly moving averages, the trend of the series;

   (ii) calculate the seasonal variations.

(c) State the main factors affecting a time series.

11.5 The following data gives the index numbers for cost of labour for a small firm during the period 19-5-19-8. By calculating an appropriate moving average obtain the trend and hence the seasonal factors for each quarter. Explain how to obtain a forecast of the index for the fourth quarter of 19-8.

Year	Quarters			
	1	2	3	4
19-5				103.5
19-6	102.2	101.2	95.1	103.5
19-7	99.1	103.8	96.8	109.0
19-8	112.1	110.5	105.2	

11.6 Supplies and Deliveries of Crude Steel
(Thousand Tons)

Year	Quarters			
	1	2	3	4
19-5	480	520	466	526
19-6	541	541	440	510
19-7	538	544	493	531

(a) Calculate the average seasonal variation for this data.
(b) Deseasonalise the series.
(c) Briefly comment on your findings.

11.7

Year	Quarters			
	1	2	3	4
19-0		110	120	140
19-1	110	124	130	154
19-2	123	134	143	176
19-3	139			

The above data gives the index numbers for the total value of sales in catering in Great Britain for the years 19-0-19-3. By calculating an appropriate moving average obtain the trend and hence the seasonal factors for each quarter. Deseasonalise the quarters of 19-2.

11.8 Days lost due to Industrial Accidents

	Quarters			
	1	2	3	4
19-4	152	204	134	213
19-5	187	198	311	299
19-6	210	257	231	323
19-7	201	225	236	318
19-8	180			

From the data given in the table calculate the average seasonal variation in days lost in the years 19-4 to 19-7 inclusive. Hence predict the number of days lost in the first quarter of 19-8. For what reasons might your prediction differ from the actual number of days lost?

C.I.P.F.A.

11.9

	Quarters			
	1	2	3	4
19-2	58	85	97	73
19-3	64	96	107	89
19-4	76	102	115	94

The above table shows the number of visitors (in hundreds) to an hotel during a period of three years. By using the method of moving averages rewrite the series without seasonal variations. Estimate as accurately as you can the quarterly figures for 19-5 and say how reliable you think your estimates are (use the multiplicative model).

11.10 The following table gives the sales (in hundreds) of components manufactured by Company A.

Quarterly Sales in Hundreds

	Quarter			
	1	2	3	4
19-2				13
19-3	14	16	9	14
19-4	16	17	12	17
19-5	18	20	13	

Calculate the seasonal fluctuations and rewrite the series with the seasonal variations removed. Draw a graph of the trend values and project it into 19-6. Hence, or otherwise, calculate the expected sales for the next four quarters (use the multiplicative model).

11.11 Quarterly production at a paper making plant was reported as follows:

Quarterly Production (000 tonnes)

	1	2	3	4
19-2	38.8	41.3	39.0	45.6
19-3	44.7	45.2	42.0	49.9
19-4	46.7	48.2	44.5	51.3
19-5	50.1	54.6		

Using the method of moving averages, find the average seasonal deviations, and thus estimate the production figures for the last two quarters of 19-5.

11.12 What do you understand by adaptive forecasting? Give an example of its use.

11.13 Use the exponential smoothing technique on the data in 11.10. Take alpha value of 0.3, 0.5 and 0.7. Which value of alpha would you recommend?

11.14 Suppose the trend line for the data in 11.9 is given by the equation Trend = 70 + 3t. Use this equation to predict the quarterly figures for 19-5 using both the additive and multiplicative models.

Chapter Twelve

Index Numbers

Without doubt, the use of index numbers affects all of us: the Index of Retail Prices is not merely an indicator of our economic welfare, it can also affect the policies pursued by governments. After all, the Index of Retail Prices is the prime measure of trends in inflation, and so can profoundly affect the government's attitude towards its own level of spending, and its attitudes towards the labour market. Again, the Index of Production may well be used as a basis for negotiating productivity agreements. On a more particular level, we should notice that in recent times index numbers have been used increasingly by accountants. This is because accountants have realised that if their budgets are to be meaningful, then they must take inflation into account. It is absolutely vital, then, for us to appreciate just what an index is, how it is calculated, and what are its limitations. Otherwise, our knowledge of the important field of inflation accounting will be imperfect.

An Expenditure Index

An index number is merely a device to measure the change in some economic variable over time. One particular time period is chosen (called the *base* period) and the variable for that period is given an arbitrary value of 100. An *index* is then calculated for the remaining periods on the assumption that the base period has a value of 100. An index, then, gives the percentage change that has occured since that base period.

Example 1:

Al Coholic throws a (rather unusual) party each Christmas for his friends. Details of his expenditure on food and drink are as follows

	1978		1979	
	price	quantity	price	quantity
	p_0	q_0	p_n	q_n
Wine (per bottle)	£1.00	40	£1.15	50
Pork pies (each)	£0.20	100	£0.27	90
Christmas Cake (each)	£2.00	1	£2.20	1

In 1978, Al spent £1.00 × 40 + £0.20 × 100 + £2.00 × 1 = £62.00 and in 1979 he spent £1.15 × 50 + £0.27 × 90 + £2.20 × 1 = £84.00. If we call expenditure in 1978 100, then expenditure in 1979 would be

$$\frac{84}{62} \times 100 = 135.4$$

which shows that his expenditure in 1979 was 35.4% higher than in 1978. As you might expect, the index we have calculated is called an *expenditure index*. If we call base year prices and quantities p_0 and q_0, and current year prices and quantities p_n and q_n, then we can calculate an expenditure index like this:

$$\frac{\Sigma p_n q_n}{\Sigma p_0 q_0} \times 100$$

A Price Index

Now one reason for the change in expenditure incurred by Al Coholic is the change in the price of the goods he has bought. It would be interesting to construct an index number to measure the extent of the change in price. Now expenditure is made up of two components, price and quantities bought, and any change in either (or both) of these components will cause a change in expenditure. So it must follow that if we hold quantities constant, then any change in expenditure must be due to price changes.

We know that the total expenditure incurred by Al in 1978 was
$$1.00 \times 40 + 0.20 \times 100 + 2.00 \times 1 = £62$$
If Al had bought the same quantities in 1979, then his total expenditure would have been:
$$1.15 \times 40 + 0.27 \times 100 + 2.20 \times 1 = £75.20$$
As we have held the quantities constant, the change in outlay must be entirely due to the change in price. If we call the price index for 1978 100, then the index for 1979 is

$$\frac{75.20}{62} \times 100 = 121.3,$$

i.e. prices have increased by 21.3%

Using the same symbols as before, we can calculate this index by taking

$$\frac{\Sigma p_n q_0}{\Sigma p_0 q_0} \times 100$$

It shows what the cost of goods in the nth year would be, assuming that we bought the same quantities as in the base year, and assuming that we call the base year price 100. An index calculated in this way is called *Laspeyre's Index*.

Now you may well complain that we have made a bit of a meal of calculating a price index. We could argue that the cost of 'one of each' in 1978 was £1.00 + £0.20 = £3.20, and the cost of 'one of each' in 1979 was £1.15 + £0.27 + £2.20 = £3.62. With 1978 as base year, the price index for 1979 is

$$\frac{3.62}{3.20} \times 100 = 113.1$$

An index calculated in this way is called a *simple aggregative index* and it has two major faults. Firstly, the value of the index will depend upon the pricing unit used — if we had quoted the price of wine as '£12 per case of twelve' for 1978 and '£13.80 per case of twelve' for 1979, we would have obtained a very different answer — even though we have not changed the price per bottle. The second drawback of this method is that it assumes that the items bought have the same significance to us, as it takes no account of the quantities bought. Surely, a 50% increase in the price of wine will have more significance to Al than a 50% increase in the price of Christmas cake? After all, he buys 40 bottles of wine, but only one Christmas cake! Fortunately, Laspeyre's Index suffers from neither of these defects. By taking into account the quantities bought, it recognises that some items bought have more significance to us than others. Using the statistician's jargon, Laspeyre's Index gives a *weighting* to the individual items to show their relative importance, using the quantities bought in the base year as weights.

Price Relatives

Before we judge the efficiency of Laspeyre's Index we should note that in practice it is not usually calculated in the way we have done — it is more usual to use *price relatives*. Price relatives refer to individual items in the index, and simply show the change in price per item since the base year. So the price relatives are calculated like this

$$\frac{p_n}{p_0} \times 100$$

The price relatives for the items bought by Al Coholic would be

	Price in 1978 p_0	Price in 1979 p_n	Price Relative
Wine	£1.00	£1.15	115.0
Pork Pies	£0.20	£0.27	135.0
Christmas Cake	£2.00	£2.20	110.0

To calculate the price index, we must now weight the price relatives, and the appropriate weights for Laspreyre's Index is expenditure in the base year.

	1978 price p_0	quantity q_0	expenditure (weight)	Relative	Weight × PR
Wine	£1.00	40	40	115.0	4600
Pork Pies	£0.20	100	20	135.0	2700
Christmas Cake	£2.00	1	2	110.0	220
			62		7520

The price index is now found by taking

$$\frac{\Sigma(\text{price relative} \times \text{weight})}{\Sigma(\text{weights})} \times 100 = \frac{7520}{62} = 121.3$$

So we have a choice of two methods for calculating Laspeyre's Index. If we wish to use base year quantities as weights, then we must use actual prices to calculate our price index. However, if we use base year expenditures as weights, then we use price relatives to calculate the index. As stated earlier, it is more usual in practise to use expenditure as weights. The reason is quite simply that it is easier to obtain data on expenditure (cost of living weights are obtained by the Survey of Household Expenditure — that is by sampling). Also, we can consider occasions when 'quantities' wouldn't make sense — how can one define 'quantities' of public transport? However, it would be quite possible to obtain details of personal expenditure on public transport.

Let us summarise what we have done so far. We have calculated an index called Laspeyre's Index, which tells us what we would have paid in year n for a collection of goods assuming we bought base year quantities. This description is true whether we use expenditure or quantities as weights. The point is that Laspeyre's Index uses base year statistics as weights. Now many statisticians doubt the validity of taking base year weights, as it implies that the quantities we buy do not vary over time. In many cases this will not be true. Suppose we have cost of living index based on 1970, and one of the items included is potatoes. The index for 1976 would assume that we bought the same quantity of potatoes as in 1970. But the period 1975-76 saw a dramatic increase in the price of potatoes, and one consequence of this was a fall in demand. Without doubt, people consumed considerably fewer potatoes and turned to substitutes, so because of this fall in quantity purchased the Laspeyre index tends to *overstate* increases in price. Another problem is that tastes tend to change fairly markedly over time, so weights that were appropriate for 1970 will not be appropriate for today. For example, few colour television sets were bought in 1970 in preference to monochrome, but today this situation is completely reversed.

In an attempt to overcome these problems, many statisticians have suggested that current year quantities should be used as weights. Our index now becomes

$$\frac{\Sigma p_n q_n}{\Sigma p_0 q_n} \times 100$$

and such an index is called a *Paasche index*. The approach here is quite different: we ask what would be the total outlay in the base year if we bought current year quantities, and compare this with current year outlay. Paasche's Price Index for the data supplied by Al Coholic is

	1978	1979			
	p_0	p_n	q_n	$p_0 q_n$	$p_n q_n$
Wine	£1.00	£1.15	50	50	57.5
Pork Pies	£0.20	£0.27	90	18	24.3
Christmas Cake	£2.00	£2.20	1	2	2.2
				70	84.0

$$\text{Paasche's Price Index} = \frac{84}{70} \times 100 = 120$$

This certainly overcomes the main objections to the Laspeyre Index, but does tend to raise problems of its own. Firstly, the Paasche index is not a pure price index as it also takes into account changes in quantities bought. Just, then, as the Laspeyre index tends to overstate the effect of rising prices, the Paasche index tends to understate the effects. Secondly, it can be a long and expensive job calculating the values for weights with a Paasche index. This must be done for every period, while with a Laspeyre index this is done for the base year only. The current cost of living index is a compromise between the two: the index is published monthly and the weights are adjusted annually by the Survey of Consumer Expenditure.

Volume Index

So far, we have assumed that the change in expenditure incurred by Al Coholic has been due to price changes, and we have attempted to measure the extent of the price change with a price index. However, the change in expenditure could be due to changes in quantities bought, and we can measure the extent of these changes by a *volume index*. We have been comparing Al Coholic's expenditure at two moments of time. If we keep the quantities bought the same, then any variation in outlay must be due to changes in price, so we can calculate a price index. If, however, we keep the prices the same, then any variation in outlay must be due to changes in quantities bought, so we can calculate a volume index. If, then, we use quantities as weights we obtain a price index, but if we use prices as weights then we obtain a volume index. Again, we have the choice between current year and base year weighting, so we have

$$\text{Laspeyre's Volume Index} \quad \frac{\Sigma p_0 q_n}{\Sigma p_0 q_0} \times 100$$

$$\text{Paasche's Volume Index} \quad \frac{\Sigma p_n q_n}{\Sigma p_n q_0} \times 100$$

Let us now calculate both indexes for the data supplied by Al Coholic.

	1978 Price	1978 Quantity	1979 Price	1979 Quantity	$p_0 q_0$	$p_0 q_n$	$p_n q_0$	$p_n q_n$
	p_0	q_0	p_n	q_n				
Wine	1.00	40	1.15	50	40	50	46.0	57.5
Pork Pies	0.20	100	0.27	90	20	18	27.0	24.3
Christmas Cake	2.00	1	2.20	1	2	2	2.2	2.2
					62	70	75.2	84.0

$$\text{Laspeyre's Volume Index} = \frac{70}{62} \times 100 = 112.9$$

$$\text{Paasche's Volume Index} = \frac{84.0}{75.2} \times 100 = 111.7$$

Selecting a Suitable Base Period

One of the main problems in constructing index numbers is the choice of a suitable base period. We wish really to choose a base when the price is as 'normal' as possible, i.e. when the price is not unduly high or unduly low. Otherwise, the index will move away from the base figure too quickly and show very large deviations from it. Suppose, for example, the price of a particular stock at certain periods of time was

April   125p   (takeover rumour)
July   69p   (takeover unsuccessful, poor dividends announced)
Dec.   95p   (quite good dividends forecast)

then the price index for this particular stock could be

April = 100.		July = 100		Dec. = 100	
April	100		181.2		131.6
July	55.2		100		72.6
Dec.	76.		137.7		100.

The best base to choose is probably December, because this minimises the greatest deviation from the base (31.6). Does this matter? Well, there is evidence to show that people are more likely to understand and appreciate smaller percentage changes than larger ones. This is one of the reasons that statisticians tend to update the bases that they use. Another reason is that bases in the not-too-distant past tend to be much more meaningful to the users of the index. It would be much more reasonable to compare prices now with prices ten years ago than it would be to compare them with prices in 1949.

This problem of choosing a suitable base can also be illustrated if we are constructing an index to measure the volume of production. If we take as our base a month in which there is a major strike, then the index in the following months would be bound to show a substantial increase and give a misleading impression of the prosperity of the industry. If we were constructing an index of motor-car sales, which do you think would be the *worst* month to choose as a base?

An alternative to up-dating the base period is to use instead a *chain-based system*, where the base used is the previous period. This method has found particular favour in the United States. To illustrate the difference between the chain-based and fixed base system, consider the following example which refers to the average price of a certain security over six months.

Month	Price(p)	Price index (chain-based, previous month = 100)	Price index (fixed base, August = 100)
August	155		$\frac{155}{155} \times 100 = 100$
September	143	$\frac{143}{155} \times 100 = 92.3$	$\frac{143}{155} \times 100 = 92.3$
October	144	$\frac{144}{143} \times 100 = 100.7$	$\frac{144}{155} \times 100 = 92.9$

November	139	$\frac{139}{144} \times 100 = 96.5$		$\frac{139}{155} \times 100 = 89.7$	
December	140	$\frac{140}{139} \times 100 = 100.7$		$\frac{140}{155} \times 100 = 90.3$	
January	131	$\frac{131}{140} \times 100 = 93.6$		$\frac{131}{155} \times 100 = 84.5$	

Consider the price in (say) January. Using the chain base numbers, we can see that the price was 6.4% lower than the previous month and using the fixed base numbers we can see that the price was 15.5% lower than in August. We can see, then, that chain based index numbers are particularly suited for period-by-period comparisons, but if we are to compare the movement of prices over time then the fixed based indexes are much easier to interpret.

An 'Ideal' Index

Earlier, we stated that Laspeyre's Index tends to overstate and Paasche's Index tends to understate changes in prices or quantities. So we would expect Laspeyre's Index to exceed Paasche's Index, and if we examine the indexes of Al Coholic's expenditure, we find that this is indeed the case.

	Volume	Price	Expenditure
Laspeyre's	112.9	121.3	
Paasche's	111.7	120.0	135.4

However, you must not assume that Laspeyre's Index will always be greater than Paasche's Index. (In fact, you should consider what the implications would be if Paasche's Index is greater than Laspeyre's Index). What is reasonable, however, is to conclude that the 'true' index lies somewhere between Laspeyre's Index and Paasche's Index, and Fisher has suggested that it is equivalent to the geometric mean of the two i.e.

Fisher's Index = $\sqrt{\text{(Laspeyre's Index)} \times \text{(Paasche's Index)}}$.

Fisher's Price Index = $\sqrt{121.3 \times 120} = 120.6$

Fisher's Volume Index = $\sqrt{112.9 \times 111.7} = 112.3$

It is claimed that Fisher's is an 'ideal' index and it now remains to state in what sense it is ideal. At the begining of this chapter, we stated that expenditure was made up of prices and quantities, so if we have a price index and a quantity index we should be able to correctly predict the expenditure index. Remembering that expenditure is price times quantities, the prediction for the expenditure index is

(a) Using Laspeyre's data,

$$\frac{112.9 \times 121.3}{100} = 136.9$$

(which overstates the expenditure index)

(b) Using Paasche's data,

$$\frac{111.7 \times 120.0}{100} = 134.0$$

(which understates the expenditure index)

(c) Using Fisher's data,
$$\frac{120.6 \times 112.3}{100} = 135.4$$

Fisher's Index numbers are ideal in the sense that they are the only ones that correctly predict the expenditure index.

Using an Index for 'Deflating' a Series

It is a well-known fact that wages have risen dramatically since 1960, and standards of living have risen too. However, as we all know, price rises can erode increases in earnings, and economists have coined the phrase 'real wages' which shows how incomes have changed, taking price changes into account. Let us examine the relevant statistics.

Year	Average weekly earnings of Manual Workers in Manufacturing Industries (£)	Retail Price Index
1960	15.16	114.5
1961	15.89	117.5
1962	16.34	100.0
1963	17.29	103.6
1964	18.66	107.0
1965	20.16	112.1
1966	20.78	116.5
1967	21.89	119.4
1968	23.62	125.0
1969	25.54	131.8
1970	28.91	140.2
1971	31.37	153.4
1972	36.20	164.3
1973	41.52	179.4
1974	49.12	191.8

Source: Annual Abstract of Statistics.

The first problem that we notice is that the price index has undergone a change of base. Earlier figures have 1956 as base, while later figures have 1962 as base. We will need to obtain a price index related to a single base, and it is usual to take the most recent base as the base for the entire series. So we will want to recalculate price indexes for 1960 and 1961 with 1962 as base. Now when a base changes, it is usual to calculate a price index for a few periods using both the new and the old base. Consulting the Annual Abstract, we see that for 1962 the price index using the old base was 119.3, so we have:

$$\text{Price index for 1960} = 114.5 \times \frac{100}{119.3} = 95.98$$

$$\text{Price index for 1961} = 117.5 \times \frac{100}{119.3} = 98.49$$

We can now use the retail price index to deflate actual earnings and obtain real wages (taking into account price rises since 1962) For example, average wage for 1970 at 1962 prices is

$$\frac{28.91}{140.2} \times 100 = £20.62$$

So actual and real wages over the period would look like this:

Year	Average Weekly Earnings (£)	Real Income
1960	£15.16	£15.79
1961	£15.89	£16.13
1962	£16.34	£16.34
1963	£17.29	£16.69
1964	£18.66	£17.44
1965	£20.16	£17.98
1966	£20.78	£17.84
1967	£21.89	£18.33
1968	£23.62	£18.90
1969	£25.54	£19.38
1970	£28.91	£20.62
1971	£31.37	£20.45
1972	£36.20	£22.03
1973	£41.52	£23.14
1974	£49.12	£25.61

When actual weekly earnings are deflated, a very different picture emerges; though without doubt living standards have risen. Notice that in all years except 1971 wages were rising faster than prices. We could if we wish calculate an index of real wages with 1962 (or any other year for that matter) as base by dividing current year real wages by 16.34 and multiplying by 100. So the index for 1970 would be:

$$\frac{20.62}{16.34} \times 100 = 126.2$$

Exercises to Chapter 12

12.1 What is the main purpose of an index number? Describe the main methods for constructing index numbers, indicating the advantages and disadvantages of each. C.I.P.F.A.

12.2 The following table shows the U.S. consumption (millions of pounds) and price (dollars per pound) of vegetable oil products

	19-3		19-4		19-5	
	Quantity	Price	Quantity	Price	Quantity	Price
Soybean	322	0.13	368	0.12	367	0.13
Cotton seed	96	0.15	114	0.14	123	0.14
Linseed	32	0.13	31	0.13	19	0.12

Required: Taking 19-3 as the base year, calculate Laspeyre's base weighted index numbers for the general level of prices of these products for 19-4 and 19-5. A.C.A.

12.3　Recalculate Laspeyre's index numbers for the data in example 2 but this time use price relatives and make 19-4 the base year.

12.4　A steel stockist notices that prices and values of sales for the main units of steel supply were:

	19-9		19-4	
	Price per tonne £	Sales £m	Price per tonne £	Sales £m
Ingots	162	324	200	600
Steel bars	188	564	190	760
Steel strip	220	880	275	1100

Calculate a Paasche index number for 19-4 prices, taking 19-9 = 100

Recalculate the Paasche price index using price relatives and the appropriate weights.

I.C.S.A.

12.5　A company employs three grades of male direct operators, M1, M2 and M3, and three grades of direct female operators, F1, F2, and F3. The following table gives the number of operators employed and the rates paid.

Labour Grade	19-1		19-3	
	rate per hour	No. of Ops.	rate per hour	No. of Ops.
M1	0.66	32	0.80	35
M2	0.62	14	0.74	10
M3	0.56	16	0.66	12
F1	0.44	40	0.64	45
F2	0.41	18	0.58	16
F3	0.36	25	0.52	30

With 19-1 as base calculate an index number for the average wage in 19-3 using

(a)　base weighting, and
(b)　current weighting

Comment on the relationship between the two indices.

12.6　The prices and sales by value of four commodities are shown below for 19-8 and 19-2. Calculate Laspeyre's and Paasche's index numbers for 19-2 taking 19-8 = 100. Would you expect the two answers to have the same value?

Commodity	19-8		19-2	
	Price £	Sales by value £m	Price £	Sales by value £m
A	2.00	8.0	3.00	12.0
B	1.00	6.5	1.50	9.0
C	4.50	9.0	5.50	11.0
D	7.00	14.0	8.50	17.0

I.C.S.A.

12.7 The following data shows the price index numbers of three groups of commodities for five years. Groups 1 and 2 have been prepared on the fixed base principle, but group 3 numbers are chain based and unlinked.

Year	1	2	3	4	5
Group 1	100	106	113	122	128
Group 2	84	94	100	108	114
Group 3	100	102	104	101	103

(a) Recalculate the Group 2 numbers to make them more easily comparable with the group 1 numbers.

(b) Convert the group 3 numbers into a series linked to the index number of 100 for year 1.

(c) Tabulate the figures for group 1 and the revised figures for Groups 2 and 3 and briefly comment on the situation revealed.

12.8 Imports Through Certain U.K. ports

	Declared Value £ million	Value at 19-0 prices £ million
19-0	522	522
19-1	430	533
19-2	351	469

Construct index numbers for price and volume with 19-0 as base year.

12.9
	Laspeyre's	Paasche's	Fisher's
Price Index	—	—	107.4
Volume Index	—	108.5	—

Find the missing data given that the Expenditure Index is 115.7, and given that base year expenditure at base year prices is £58,430 million.

12.10
Year	Profit (£000)	Wholesale price index
1970	1132	112.3
1971	1245	120.7
1972	1345	129.3
1973	1412	100.0
1974	1582	110.4
1975	1898	127.1
1976	2182	147.4
1977	2488	162.1
1978	2736	175.1
1979	2955	183.9

Before 1973, the wholesale price index had 1965 as its base year. Index for 1973 with 1965 = 100 is 136.5.

(a) Use the index of wholesale prices to deflate profits, and so obtain profits at 1973 prices.

(b) Find the index numbers of 'real profits' for 1970-1979 with 1973 = 100.

Chapter Thirteen

Dispersion

In Chapter 6 we found it desirable to summarise data with a measure of central tendency and the main reason we did this was to enable us to compare sets of data or frequency distributions. We spent quite some time deciding how such measures could be truly representative of the data, and discussed the suitability of using the arithmetic mean (or average), the median and the mode. However, it is not sufficient to summarise data with a measure of central tendency only, we also need a measure of dispersion. The problem with quoting a measure of central tendency only is that we can be led into drawing the wrong conclusions. Suppose, for example, we were examining income at two periods of time, and we find that there has been no change in average income. We might be tempted to conclude that economic welfare has not changed. However, although average income may not have changed, it is quite possible that welfare did increase through a more equal distribution of income. While our measure of central tendency will detect changes in average income, a measure of dispersion will detect changes in the spread of income. Clearly, both measures are desirable, so we shall now critically examine methods of measuring dispersion.

Measures of the Range

To measure dispersion, we could simply take the difference between the greatest and the least value in a series or distribution. We call this measure the *range*. To illustrate how the range is calculated, consider the data below which gives the average price of a certain security over a six month period.

	Average Price (p)
Jan.	155
Feb.	143
March	144
April	139
May	140
June	141

So the highest price during the period was 155p and the lowest price was 139p, giving a range of 16p. Nothing, then, could be more simple both in concept and calculation. However, the range does have two great faults. Firstly, it ignores the bulk of the data available to us, being concerned only with extreme values. The second disadvantage is that, being concerned with the extreme values only, the range may be quite unrepresentative of the spread of items, especially as the extreme values are probably quite untypical of the data. Suppose, for example, we were to use the range as a measure of dispersion of the heights of males in Liverpool. The range would be the difference between the tallest giant (say 6 foot 8 inches) and the smallest dwarf (say 4 foot 6 inches) — a range of 26 inches. However, we would probably find that by far the bulk of the population would be contained within the limits 5 foot 6 inches to 6 foot 2 inches — a range of 8 inches. Despite these criticisms, the range is used considerably — especially in statistical quality control. The reason is that in this field a measure of dispersion is often required quite quickly, and the range is eminently suited to this condition. However, it should be pointed out that, in these circumstances, the range will be found repeatedly from a large number of samples, and this does minimise the chances of obtaining untypical values at the extremes.

Because the range can be so misleading, statisticians have turned to alternative measures of dispersion. The one we shall now examine finds the range containing the central 50% of a distribution. Consider the data below which refers to the bonuses paid to the salesmen of a particular firm during a particular month.

Monthly bonus (£)	Number of salesmen	Cumulative Frequency
Under 60	8	8
60 and under 70	10	18
70 –	16	34
80 –	14	48
90 –	10	58
100 and under 110	5	63
110 and over	2	65
	65	

We want to find the range of bonuses earned by the central 50% of salesmen. We can do this by finding the bonus earned by the $\frac{65+1}{4} = 16.5$th salesmen and the bonus earned by the $\frac{3(65+1)}{4} = 49.5$th salesman. Diagram 13.01 shows the relevant part of the ogive from which we can read that 16.5 salesmen earn a bonus of less than £68.50 and 49.5 salesmen earn a bonus of less than £91.50. So the range earned by the central 50% of salesmen is £91.50 — £68.50 = £23. We call this range the *interquartile range* (it is the range of bonuses earned by the central two quarters of the distribution). We call the lower end of this range (£68.50) the *lower quartile (Q_1)* and 25% of salesmen earn less than this figure. Likewise, the upper end of the range is called the *upper quartile* (Q_3) and 25% of salesmen earn more than this figure. Calling the lower quartile Q_1 and the upper quartile Q_3 would naturally lead us to call the median Q_2. Notice also that although the median *salesman* lies midway between the upper quartile salesman and the lower quartile salesman, the median *bonus* is not at the centre of the interquartile range. Beware of this — many students fall into this trap.

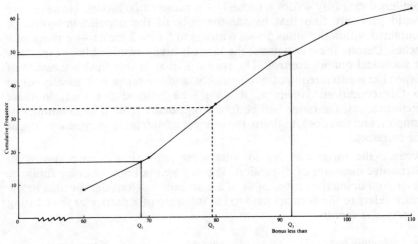

Diagram 13.01

Sometimes you will come across the *quartile deviation* — this is merely half the interquartile range, i.e.

$$\text{Quartile deviation} = \frac{Q_2 - Q_1}{2}$$

so the quartile deviation for monthly bonuses is $\frac{23}{2}$ = £11.50.

Notice that in this example, we could not have calculated the range for bonuses, as we have no way of knowing the highest bonus nor the lowest bonus earned. The range cannot be used for open-ended distributions. This disadvantage will probably not apply to the interquartile range, as it is unlikely that the central 50% of frequencies will penetrate the open ended class. However, the interquartile range does have the same disadvantage as the range — it uses only two values and ignores the rest. What we shall now examine are measures of dispersion that use all the data available to us.

Measures of Average Deviation

So far, we have examined two predetermined points in a series or distribution and calculated the difference between them. We have fixed the points either at the ends of the distribution (the range) or about the central 50% (the quartile deviation). We could of course, have fixed these two points anywhere we wished; for example, covering the central 40% or 60% of the items. But let us now consider dispersion in an entirely different way: we could choose some central value and calculate the deviation of the items from this value. What central value should we choose? Well, the arithmetic mean is one choice that automatically springs to mind. However, there is a snag with choosing the arithmetic mean — can you see what it is? Clearly, we are not going to be interested in the individual deviations themselves — we would still have a series of numbers rather than a single measure of dispersion. Surely, it would be more sensible to consider the average deviation from the mean. Now here's where the snag comes in. If you remember the definition of the arithmetic mean, then you will appreciate that the sum of the deviations from the mean is zero. Hence the average deviation from the mean must always be zero! Clearly, this will not do — we cannot use the average deviation from the mean as a measure of dispersion, because irrespective of the distribution under consideration our measure of dispersion will always be zero.

One way out of this problem is to find the *absolute deviations* from the mean, irrespective of whether they are positive or negative. Let us do this for the data on security prices considered earlier. The average price was

$$\frac{155 + 143 + 144 + 139 + 140 + 141}{6} = \frac{862}{6}$$

$= 143\tfrac{2}{3}\text{p}$

and the deviations from this price are

Price (x)	Deviation $x - \bar{x}$	Absolute Deviation
155	11⅓	11⅓
143	−⅔	⅔
144	⅓	⅓
139	−4⅔	4⅔
140	−3⅔	3⅔
141	−2⅔	2⅔
	0	23⅓

So the average absolute deviation from the mean is

$$\frac{23\tfrac{1}{3}}{6} = 3.89$$

The measure of dispersion we have just calculated is called the *mean absolute deviation,* and using the Σ notation we calculate it like this:

$$\text{Mean absolute deviation} = \frac{\Sigma |x - \bar{x}|}{M}$$

The vertical lines enclosing $x - \bar{x}$ simply means, "take the absolute deviations", whereas brackets would imply that we must take account of signs. Only a minor adjustment is needed to this formula if we are confronted with a frequency distribution.

$$\text{Mean absolute deviation} = \frac{\Sigma (x - \bar{x})}{\Sigma f}$$

where f represents the frequencies and x represents the centre point of the various classes.

The mean absolute deviation has a number of disadvantages. Because we have taken absolute deviations and ignored the signs, it is not capable of being manipulated mathematically in the ways that the statistician requires. A second disadvantage is that it will be quite cumbersome to calculate when the mean is not a whole number. Thirdly, we cannot combine a number of mean absolute deviations to obtain an overall measure of dispersion. Suppose, for example, a number of samples will be made available to us over a period of time, and we intend to use the samples to estimate dispersion in the population. We cannot calculate the mean deviation of each sample, and pool them, adjusting our estimate as more samples become available. We would have to pool all the samples, making one single sample. We do not wish, however, to leave you with the impression that the mean absolute deviation is of little use — far from it. One thing it cannot be criticised for is its ability to be a *representative* measure of dispersion. Probably no other measure of dispersion does this quite so well. The advantage of the mean absolute deviation is that conceptually it is so easy to understand. So what advice can we give you about when to use it? If you are interested merely in representing dispersion, then you can do no better than to use the mean absolute deviation, but if further statistical analysis is required we must look to other measures.

The main problems surrounding the mean absolute deviation arise because we took all the deviations as being positive in order to prevent the sum of the deviations being zero. Is there any other course of action open to us? Well, we could square all the deviations — the square of a negative number is a positive number, so the pluses and minuses will no longer cancel each other out. If we find the average of the square of the deviations, then we have a measure called the *variance*.

$$\text{Variance} = \frac{\Sigma(x - \bar{x})^2}{n}$$

So we would calculate the variance of security prices as follows:

Price (x)	$(x - \bar{x})$	$(x - \bar{x})^2$
155	11.33	128.44
143	−0.67	0.44
144	0.33	0.11
139	−4.67	21.78
140	−3.67	13.44
141	−2.67	7.11
		171.32

$$\text{Variance} = \frac{171.32}{6} = 28.55p$$

The trouble with the variance is that (in this example) it is measured in units of the square of the price deviations. Strictly speaking, the variance of security prices is not 28.55p but 28.55 (pence)². It would seem more sensible, then, to take the square root of the variance. We give the name *standard deviation (σ)* to this measure of dispersion, and of all the measures this is the one that is most commonly used.

$$\sigma = \sqrt{\frac{\Sigma(x - \bar{x})^2}{n}}$$

So the standard deviation of the price of securities is $\sqrt{28.55} = 5.34$p.

It is quite easy so show that an alternative form of the standard deviation is

$$\sigma = \sqrt{\frac{\Sigma x^2}{n} - \left(\frac{\Sigma x}{n}\right)^2}$$

Notice that this expression does not ask us to calculate the deviations from the arithmetic mean — and this will be very useful when the mean is an awkward number. We do not intend to prove that the two expressions are the same (you can do this for yourself) but we shall demonstrate that both expression give the same result by calculating the standard deviation of the price of securities this time using the second expression.

x	x^2
155	24,025
143	20,449
144	20,736
139	19,321
140	19,600
141	19,881
862	124,012

$$\sigma = \sqrt{\frac{124012}{6} - \left(\frac{862}{6}\right)^2} = 5.34\text{p}$$

If we wish to find the standard deviation of a frequency distribution, we simply replace x with fx and n with f.

$$\sigma = \sqrt{\frac{\Sigma f x^2}{\Sigma f} - \left(\frac{\Sigma f x}{\Sigma f}\right)^2}$$

where x is the centre point of the class and f is the frequency. Let us now look at an example.

ABC Ltd. Debtors' Balances at 31st December 19-9

Balance Outstanding (£)	Number of Accounts
20 and under 40	1
40 and under 60	3
60 and under 80	6
80 and under 100	10
100 and under 120	5
120 and under 140	3
140 and under 160	2
	30

We need first to find the centre point of each class.

Centre point (x)	f	fx	$fx^2 = fx \times x$
30	1	30	900
50	3	150	7500
70	6	420	29400
90	10	900	81000
110	5	550	60500
130	3	390	50700
150	2	300	45000
$\Sigma f = 30$		$\Sigma fx = 2740$	$\Sigma fx^2 = 275000$

$$\sigma = \sqrt{\frac{275,000}{30} - \left(\frac{2740}{30}\right)^2}$$

$$= £28.72$$

Notice that in constructing the table for calculating the standard deviation, we have also obtained all the information we need to calculate the arithmetic mean. It would make sense, then to calculate the mean as well as the standard deviation.

$$\bar{x} = \frac{\Sigma fx}{\Sigma f} = \frac{2740}{30} = £91.33$$

Unless you have a calculator at your disposal, it can be very tedious to calculate the standard deviation using this method. Fortunately we can simplify the calculations (you will remember that we did this when calculating the arithmetic mean) by taking an assumed mean — in this case we will take £90. We then calculate the deviation of the centre points x from the assumed mean (column 2). As all the deviations are divisible by 20, this is done in the column headed d. We can now use the column d to calculate the standard deviation.

x	(x − 90)	d	f	fd	fd^2
30	−60	−3	1	−3	9
50	−40	−2	3	−6	12
70	−20	−1	6	−6	6
90	0	0	10	0	0
110	20	1	5	5	5
130	40	2	3	6	12
150	60	3	2	6	18
			$\Sigma f = 30$	$\Sigma fd = 2$	$\Sigma fd^2 = 62$

Calculating the standard deviation as before, we have

$$\sigma = \sqrt{\frac{62}{30} - \left(\frac{2}{30}\right)^2}$$

$$= £1.436$$

However, we divided all the deviations by 20, so we must multiply the standard deviation above by 20

$$\sigma = £1.436 \times 20 = £28.72$$

which agrees exactly with our previous result. Let us see if we can write a formula which describes the above calculation. If d is the deviation from an assumed mean x_0, and if d is divided by a constant c, then

$$\sigma = c \times \sqrt{\frac{\Sigma fd^2}{\Sigma f} - \left(\frac{\Sigma fd}{\Sigma f}\right)^2}$$

Relative Dispersion

So far, we have been trying to measure dispersion within a series or distribution. We shall now attempt to measure dispersion between distributions, and we shall find that the standard deviation cannot do this. Consider the two series below:

A = 8,9,10,11,12,13,14   Mean 11
B = 1008,1009,1010,1011,1012,1013,1014   Mean 1011

Both A and B have the same standard deviation — can you see why? The deviations from the mean are identical in both series ($-3, -2, -1$ etc.) and both series have the same number of observations, so $\Sigma(x - \bar{x})^2$, and n would be the same whichever series we examined. You should verify that the standard deviation is, in fact, 2 for both series. Now does this mean that both series show the same degree of spread? Surely not! An increase from the smallest to the largest value in A is an increase of

$$\frac{14 - 8}{8} \times 100 = 75\%$$

but the same increase for series B is only

$$\frac{1014 - 1008}{1008} \times 100 = 0.595\%!$$

Clearly, A has a much greater percentage spread than B, and if we are going to compare dispersion between series we must take into account not only dispersion within each series (i.e. the standard deviation) but also the mean of the series. We can do this by calculating the *coefficient of variation*

$$V = \frac{100 \times \sigma}{\bar{x}} \%$$

For the series above, the coefficients of variation are

$$\frac{100 \times 2}{11} = 18.18\% \text{ (for A)}$$

and $$\frac{100 \times 2}{1011} = 0.198\% \text{ (for B)}$$

confirming our suspicion that A has a greater relative spread than B. Notice that the coefficient of variation is a percentage, and not a particular measurement (centimetre, dollar, kilogram, etc.) and so it can be used for comparing distributions which have different units.

Now we would like to issue a word of warning. If you attempt to quantify the differences in dispersion using the coefficient of variation, ask yourself whether the result you have obtained 'feels right'. Compare your result with a 'gut feeling' you have obtained by visually examining the histogram of each distribution. If your results do not feel right then you would be well advised to compare the distributions using the quartiles. Earlier, we defined the quartile deviation as $\frac{1}{2}(Q_3 - Q_1)$, and we could take the median Q_2 as a measure of central tendency, so an alternative measure of relative dispersion is

$$\frac{100 \times \frac{1}{2}(Q_3 - Q_1)\%}{Q_2} = \frac{50(Q_3 - Q_1)\%}{Q_2}$$

and we shall call this measure the *quartile coefficient of variation*. We shall say no more about it now, but in the assignments at the end of this chapter you will find an example where its use is obviously required.

Exercises to Chapter 13

13.1 From the figures below state what is
 (a) the range
 (b) the arithmetic mean
 (c) the median
 (d) the lower quartile
 (e) the upper quartile
 (f) the quartile deviation
 (g) the mean deviation

8	35	45	50	60	68
13	37	46	52	61	70
26	40	47	55	65	71
29	41	48	58	67	75
33					

I.C.M.A.

13.2 The following frequency distribution gives the monthly expenditure on advertising in a certain year by the branches of a large chain store

£	Number of branches
800 and less than 1000	50
1000 and less than 1200	200
1200 and less than 1400	350
1400 and less than 1600	150
1600 and less than 1800	100
1800 and less than 2000	75
2000 and less than 2200	50
2200 and less than 2400	25

Find the median and quartile deviation.

13.3   A.B.C. Ltd. Debtors Balances at 31st December 19-9

Balance Outstanding (£)	Number of accounts
20 and under 40	1
40 and under 60	3
60 and under 80	6
80 and under 100	10
100 and under 120	5
120 and under 140	3
140 and under 160	2

Find the median and quartile deviation of the debtor's balances.

13.4   The frequency distribution below gives details of the bonuses paid by a firm to its salesmen in a particular month.

Bonus paid (£)	No. of salesmen
50 and under 60	8
60 and under 70	10
70 and under 80	16
80 and under 90	14
90 and under 100	10
100 and under 110	5
110 and under 120	2

Find the mean and standard deviation.

13.5   The following frequency distribution gives the ages of people entitled to receive pensions from a firm's superannuation scheme

Age (years)	No. receiving pensions
65 – 69	1030
70 – 74	820
75 – 79	460
80 – 84	240
85 – 89	80
90 and over	20

Find the mean and standard deviation.

13.6   A furnace whose size is nominally 200 tons is used to cast steel ingots of 10 tons weight. The amount of steel in the furnace cannot be controlled accurately, so that an incomplete ingot is normally produced when ingots are cast. For example, a furnace load of 198 tons will produce 19 full ingots and one 8-ton ingot. The following data show the weight of 100 furnace loads of steel.

Weight of Furnace load (tons)	Frequency
190.0 and under 192.5	1
192.5 and under 195.0	4
195.0 and under 197.5	8
197.5 and under 200.0	19
200.0 and under 202.5	36
202.5 and under 205.0	20
205.0 and under 207.5	8
207.5 and under 210.0	4

Required: Compile a frequency distribution of the weight of incomplete ingots and calculate its mean and standard deviation.   A.C.A.

13.7 The following table shows the distribution of weekly income for two towns

Weekly Income (£)	Town A No.	Town A Income (£)	Town B No.	Town B Income (£)
0 and under 10	0	0	1	5
10 and under 20	4	60	5	75
20 and under 30	9	225	9	225
30 and under 40	17	595	13	455
40 and under 50	23	1035	19	855
50 and under 60	19	1045	23	1265
60 and under 70	13	845	17	1105
70 and under 80	9	675	9	675
80 and under 90	5	425	4	340
90 and under 100	1	95	0	0

(a) Show that the arithmetic mean income is the same in both towns. In the lights of this result, do you feel that the arithmetic mean is a useful measure to use for comparison of the two towns 'average' income? If not, derive an alternative measure that you consider better, giving reasons for your choice.

(b) Show that the standard deviation of incomes is the same in both towns. In the light of this result do you feel that the standard deviation is a useful measure to use for the comparison of the two towns' dispersion of income? If not derive an alternative measure that you consider better, giving reasons for your choice.

C.I.P.F.A.

13.8 A company has two factories A and B, situated in different parts of England. Labour turnover in factory B is much higher than in factory A. The consultant engaged by the firm to investigate the causes of labour turnover suggests that a possible cause of the higher turnover in factory B is a wider variation in the annual wages of workers in factory B than in factory A: he found that the standard deviation of wages of workers in factory B was £600 compared with a standard deviation of £500 for factory A. Given the information that the mean annual wage of workers in factory B is £2000 and in factory A £1500, show how the consultants suggestion may be tested.

A.C.A.

13.9 In department A of a certain firm, the average weekly wage is £36.60 with a standard deviation of £2.82. The weekly wages of the workers in department B are

£23.70 £24.50 £28.30 £29.60 £25.30 £24.90 £23.60
£25.40 £30.50 £24.20

Compare wages earned in the two departments.

13.10 The following data refers to weekly earnings of craftsmen in a certain firm

Weekly Earnings (£)	No. of Craftsmen
Under 72	4
72 and under 76	9
76 and under 80	16
80 and under 84	28
84 and under 88	45
88 and under 92	66
92 and under 96	85
96 and under 100	72
100 and under 104	54
104 and under 108	38
108 and under 112	27
112 and under 116	18
116 and under 120	11
120 and under 124	5
124 and over	2

For a similar group of workers in a rival firm the coefficient of variation of wages is 18%. Compare variability of earnings in the two firms.

Chapter Fourteen

Statistical Sources

There is little doubt that the second half of the twentieth century is the age when the data bank has come into its own. Data is collected on a large scale, not only by the government, but also by firms specialising in making market surveys and taking public opinion polls and by industrial firms and international organisations such as the United Nations or European Economic Community. This data is stored and made available to investigators through data retrieval systems. At one extreme we may be seeking data from a highly sophisticated computer system; at the other we will be painstakingly searching through the files in a filing cabinet.

It is small wonder that in the last half century the nature of the problem facing the statistical investigator has changed. As recently as 1920 the main problem was that data on many important problems was just not available, and the statistician had to find some means of organising his own survey to obtain primary data. Today the problem is often one of choice — of knowing where data is stored and selecting that which is most suited to immediate needs. There is still a great deal of primary data collected of course, particularly when we are concerned with marketing our product or with assessing public opinion which, at best, is fickle. Yet modern sophisticated methods of sampling still have to be based on secondary data. Even so simple a task as assessing which way the electorate will vote will require not only the register of electors, but also information on the age structure and the social structure of the areas to be sampled. The survey itself is in fact very often only the tip of the iceberg. Where then are we to look for the data which may be essential to us?

Before we start hunting through the mass of published information we must realise that much of the data we need may already be held in the records of our own firm or trade organisation. A survey into market trends could well begin with an analysis of salesmens' performance and their opinion of how the market will develop analysed by both product and area. We may not get all we require, but it will give use a good start. In the same way a great deal of useful information is held in wages books, petty cash records, ledgers, personnel records and even in the published final accounts of the firm. One of the greatest advantages of using such internal data is that it is still possible to ask questions of those who originally recorded the data, to clarify definitions, to discuss problems and possibly even to extend the data. It is just not possible to obtain this supplementary information when we are using data produced by an outside body.

At the same time we must remember that internal data is, by its very nature, narrowly based. It will relate to a single firm, or at best to a single industry. If we look at the sales analysis of a firm of seedsmen we will certainly get a very clear picture of sales trends for that firm. We may be able to extend the picture and argue that these trends are typical of the seed industry as a whole. What we cannot do is to pretend that the figures we have tell us anything about other industry or about British industry in general.

Only when we have exhausted such internal source should we turn to published data. No matter how narrow or specific our enquiry, however, we will still have to turn to secondary data because, if we are to put our results into perspective, there is still a mass of background information that we need; and often this data can be obtained only from published sources. Unemployment within our own industry becomes more meaningful when we can put it into the context of the unemployment rate within the region or within the United Kingdom; the rate of inflation means little until we compare it with the rate of increase in salaries and the rate of inflation in other countries. Much of the data we need will already have been produced by independent survey bodies such as Gallup Poll, or by research units within firms or the universities, but by far the most important source of such secondary statistics is the Central Statistical Office, responsible for publishing data collected by the wide range of government departments. Whether you want to know the average July temperature at London Airport, the number of children in primary schools last year, or the Balance of Payments in 1972, you will find it tucked away among the government statistics published by the Central Statistical Office.[1]

The Annual Abstract of Statistics

In the United Kingdom possibly the best source of this official data is the Annual Abstract of Statistics published by the Central Statistical Office. It collates and summarises information relating to the United Kingdom produced by government departments (such as the Department of Education and Science) and obtained from national surveys (such as the census of distribution or the census of population). As you might expect it tends to give information for the United Kingdom as a whole rather than individual towns or industries. If we need more detail we may have to refer to far more comprehensive reports issued by the appropriate government departments. Population statistics, for example, comprise 29 tables in the Annual Abstract, but the Office of Population Census alone publishes comprehensive and detailed reports on each individual county, together with a Preliminary Report, and a general report. Additionally, the Registrar General's Annual Review and Quarterly Reviews supplement the Census reports.

---

1. Students are urged to obtain a copy of the Stationary Office publication — "Government Statistics — a brief guide to sources" published by and issued free by The Press and Information Services, Central Statistical Office, Great George Street, London SW1P 3AQ.

The Annual Abstract is precisely what its name implies — an abstract, but it is a good starting point for any investigation. Not only does its fifteen sections cover such diverse topics as weather, output, finance, education and transport, but in most cases information is given on a comparative basis for at least ten years so that any marked trend can be spotted easily. For more recent years most data is given monthly. Additionally, the Central Statistical office publish the Monthly Digest of Statistics, bringing the information in selected fields up to date month by month.

Population and Vital Statistics

Not every statistician will agree with us but we tend to think that of all the data produced by the government, that relating to population is the most useful. General knowledge of the population and its age-sex distribution is of course necessary for efficient administration, but think of what else is needed. We need age distribution to plan schools for the next generation, to assess the effect of changes in (say) family allowances, or to plan a policy for pensions; we need a knowledge of the causes of death to allocate funds for medical research; information on housing is the first step in planning a housing policy; questions on educational qualifications help the formulation of an educational policy — and so on. Information on all these topics, and many more, can be found in the reports on the Census of Population and the Annual Reports of the Registrar General. There is little doubt that the Census of Population is one of the most useful sources of data, and certainly, along with its necessary accompaniment — the compulsory registration of births, deaths and marriages — it is the most useful single source of information relating to vital statistics.

The census consists of a complete enumeration of the persons in a country on a given night together with questions on age, sex, marital status, occupation and the industry in which they work. In addition the modern census collects information on social conditions such as the housing of the population or the educational standards reached. In the United Kingdom the first census was taken in 1801 and apart from 1941, one has been taken every 10 years since that date. The government now have, in fact, the powers to take a census every five years if they so wish, and one such was taken in 1966.

Obviously, every census must ask certain questions which are never changed — such as name, sex, age etc., but the opportunity is usually taken to ask additional questions to throw light on social problems that rear their head from time to time. In 1921, for example, questions were asked relating to widows and orphans in view of the manpower losses between 1914 and 1918; in 1961 a 10% sample was asked to answer additional questions relating to educational attainment; and in 1966 questions were asked relating to the mobility of the population, on car ownership and garage space, and on the journey to work.

While such statistics are essential, in many developing countries a complete census is not possible, probably because of the distances involved, the lack of administrative machinery for collection and analysis of data, possibly because of the literacy rate, or the knowledge that the population would not willingly give the information requried. In such cases, a limited census or even a sample survey may well be the only way to obtain the information about the population and its living conditions.

It would be foolish to pretend, however, that even in the United Kingdom the census information is completely accurate. There are sources of error at every stage. The distribution and collection of forms is in the hands of part-time enumerators who are each responsible for an area. We can never be absolutely sure that every house has received a form or that every form is returned. It is not likely, of course, but is at least possible that an enumerator may not know where a house is, and may not bother to find out, particularly if the weather is very bad, as it was on census day 1951. A much more probable source of error, however, lies in completing the questionnaire. Some people may misunderstand the question, others may only be able to guess the answer; others will believe that the answer they give is correct when, in fact, it is not. Educational standards in the U.K. vary tremendously and the quality of the completed questionnaires is bound to vary — some mistakes will inevitably occur. Admittedly some mistakes will be picked up by the enumerator on the doorstep; others will be spotted during the process of tabulation. But some will remain. How can any enumerator know that John Smith is not in fact married to the woman living in his house if he claims that he is married?

The census office itself is a fruitful source of error — such as errors arising in transferring information from the forms to punched cards or magnetic tape. That errors do exist has been shown by special post-census sample surveys designed to check accuracy, and by such techniques as comparing the death certificates of people who die shortly after the census with information given in their completed questionnaires. Yet errors are relatively few and the census office is able to claim a remarkably high degree of accuracy for most of the data obtained from the census.

Information of this nature is supplemented annually by the reports of the Registrar General in respect of births, deaths, marriages, divorces, etc., while figures for immigration and emigration can be brought up to date from Home Office statistics. In the U.K. such figures are highly reliable since there are legal pressures to ensure that registration takes place. It is not possible to bury a corpse without a death certificate, or to obtain a passport wihout a birth certificate, doctors and hospitals notify the Registrar of all births to ensure that parents ultimately register their children. Not all countries, however, have these same pressures and in some countries the births of female children are regularly not notified, while in others deaths are not registered. Sometimes of course the discrepency becomes obvious when the rate of male and female births (which stays remarkably stable) begins to change, or when the death rate differs markedly from a neighbouring country with very similar social conditions. Nor must we think that the information required is confined to the mere fact of the birth or marriage. The Population (Statistics) Acts require that informants also give such information as the age of parents, their occupation, the duration of the marriage and so on. So much information is available, in fact, that the analysis of the data has given rise to a whole new branch of statistics — demography, the analysis of vital statistics.

Economic Statistics

Since the time of the Domesday Book in the eleventh century governments have been interested in data on how the economy is progressing. In the very early days, it was, of course, merely a question of listing manpower for military purposes and personal wealth for taxation purposes. It is interesting to remember that in many underdeveloped parts of the world people are still averse to giving information for fear that it might be used to assess them for taxes. By the sixteenth century interest had developed in statistics of overseas trade since the philosophy of Mercantilism demanded a balance of payments surplus as a criterion of success. But the twentieth century is, after all, the century of economic planning, and for successful planning a mass of wide-ranging information is needed. Today government departments produce statistical data on almost every aspect of economic life — on prices and incomes, on employment and productivity, on production and distribution, on trade and finance. Once again you will find information on all these in the Annual Abstract, but let us take one or two of the major fields and look behind the statistical tables.

(a) Price Indexes

It is commonplace that whenever a body of workers ask for an increase in wages, one of the points that they make strongly is that the increase is designed to compensate for the rise in the level of prices, and they usually quote in evidence changes in the Index of Retail Prices. In Britain this index, introduced in 1947 to replace the old Cost of Living Index, is probably the best method of measuring the changes in the standard of living that we have. Constructed monthly and published in the Department of Employment Gazette it is used not only as a measure of inflationary trends, but also as the foundation of incomes policy.

If you think back to chapter twenty three you will recall that there are three main elements in the construction of a price index — the selection of commodities to be included, the determination of weight for each commodity or group of commodities, and obtaining the appropriate prices at the relevant point of time.

The present index originated in 1953 with the Household Expenditure Survey conducted by the then Ministry of Labour. This was an effort to discover the precise pattern of household expenditure in the United Kingdom. A sample of 20,000 households was selected and each person over 16 in the sample was asked to keep a detailed record of what he spent over a three-week period. The survey produced a 65% response, some 13,000 forms being returned. In the event it was found that low-income families, and particularly those living mainly on state pensions, had an expenditure pattern so different from the majority of households that they could not be regarded as typical and so were ignored. Thus the usable sample was cut down to 11,600 household budgets. It was felt that this sample reflected the spending habits of over 90% of the public and was sufficiently accurate to form the basis of a new index of retail prices. The weights to be used were determined by the percentage of income spent on individual commodities, and the survey also indicated the commodities that should be included in the index. Thus with a total weight of 1000, the weight for food was 350 and alcoholic drink 71, indicating that the average consumer spent 35% of his income on food and 7.1% of his income on alcoholic drink. On this basis a new index was introduced based on 17th January 1956 = 100. This index remained in existence for six years only, until January 1962.

Now as you know, if an index is to be really useful, the base should not be changed too frequently. Why, then, was this index dropped after only six years? It was soon realised that as wages and prices rose in the post-war world, people's consumption patterns changed fairly rapidly. Three cases in particular are outstanding: we were spending a far higher proportion of our income on such things as houses and motor-vehicles, but we were spending a much lower proportion on food. Hence it was decided that the weights used in the construction of the index should be changed year by year. This annual revision was based on a smaller Family Expenditure Survey covering a sample of 5000 households each year. These surveys are continuous throughout the year and each member of the household is required to keep a detailed account of his spending for a fortnight. It is not, of course, possible to use a single year's survey as a basis for revising the weights. Some items such as cars, carpets and washing machines are bought only infrequently, and this is likely to cause a large sampling error. So the weights are determined by the average of the previous three years' expenditure patterns disclosed by the survey. A study of the way the weights have changed over the years tells us a great deal about the way our expenditure patterns have changed.

Group	Weights used in construction of I.R.P.						
	1914	1952	1956	1962	1966	1970	1974
Food	60	399	350	319	298	255	253
Alcoholic drink		78	71	64	67	66	70
Tobacco		90	80	79	77	64	43
Housing	16	72	87	102	113	119	124
Fuel and light	8	66	55	62	64	61	52
Durable household goods		62	66	64	57	60	64
Clothing and footwear	12	98	106	98	91	86	91
Transport and vehicles			68	92	116	126	135
Services		91	58	56	56	55	54
Meals consumed away from home						43	51
Miscellaneous		44	59	64	61	65	63
Others	4						
	100	1000	1000	1000	1000	1000	1000

You can see how over the years the percentage of income spent on foods has fallen from 60% in 1914 to 25% in 1974. Since few people were starving in the U.K. in 1974 this is indicative of the rising standard of living. We can see also, signs of the revolution in power. Coal, the major fuel in 1914, absorbed 8% of income, whereas by 1974 coal and electricity absorb only 5.2%. The percentage of income spent on tobacco has also fallen rapidly, but most important of all in indicating the quality of life is the fact that if we take the four basic necessities of life — food, clothing, housing and fuel and light, we find that in 1914 90% of income was spent on these; by 1974 we spent only 52%. A study of these weights reveals a great deal about society.

Having solved the problem of obtaining a satisfactory weighting, and having made sure that the commodities covered by the index are those actually bought by the man in the street, we are still left with crucial problem of what prices to use. Everyone knows that even in a small town or village, prices vary widely from shop to shop. Hypermarkets and supermarkets sell goods at prices which smaller shops cannot possibly hope to match. Prices differ according to the brand of the commodity we buy and, of course, prices differ from area to area, reflecting partly the degree of competition in the area and partly the costs of distribution. Every motorist knows that even in his own area there may be as much as 6p difference in the price of a gallon of petrol. If he travels long distances he will also see that as he moves away from the vicinity of the oil refineries, the *general* level of prices tends to rise. The result is that while some people are paying x pence for a gallon of petrol, others may have to pay as much as ten pence more. Which of these prices are we to use when we are constructing a retail price index? The Department of Employment uses an interesting technique to try to ensure that the prices they use are representative.

On the Tuesday nearest to the 15th of the month prices are collected by visiting several shops selling the same type of good, care being taken that the types of shop visited are those that handle the bulk of household spending. This is done in several areas of different types ranging from London and the cities to small townships with a population under 5000. Thus, both different types of retail outlets and different geographical areas are represented.

Price relatives are now calculated for each item in each town. Let us take bread as an example. The price relatives for a particular city where five shop have been visited are 182.3, 181.9, 183.1, 182.5 and 182.4. These price relatives are combined with those for other cities to give a simple unweighted average. Thus we have a price relative for cities. The same analysis is now carried out for the other population groups, and the resultant price relatives averaged to give a price relative for bread for the country as a whole. In the same way we get price relatives for every commodity covered by the index.

So far, weighting has not mattered, but now we begin to combine the items to obtain an index for each of the major classes, and the question of weights becomes important. Although we know that in 1974 the class 'food' was given a weighting of 253, in order to obtain the index of food prices, this total weight has to be broken down into its constituent parts. We might, for example, apply weights of 29 to milk and cheese products, 31 to bread and cereals and so on. Once this has been done it is a simple matter to obtain an index for food, and having done the same for all other groups, to obtain an overall index of retail prices.

Can you see now why so many individuals cannot believe what the Retail Price Index tells us? As individuals we buy only a selection, and often only a small selection of the items included in the index. The items we buy may have gone up in price more than others that we do not buy, and it seems to us that the index understates the price rises. Or we may buy from a grocer whose prices have been steady when others are charging more. The index will then seem to overstate the rise in prices. Like many statistical measures, the Index of Retail Prices is a good and useful measure when applied to large numbers. It is not so useful when applied to the individual.

(b) Manpower, Production and Distribution

If you ask any economist for his 'short list' of aims for economic policy it is almost certain that among the items listed would be 'to promote economic growth'. At the risk of annoying our economist friends we will interpret this to mean simply that we are trying to increase the volume of output. Equally surely we would guess that another aim would be 'to eliminate unemployment', and a third to 'raise the standard of living'. So long as such things as these are accepted as aims of economic policy, it is little wonder that modern governments devote a great deal of time to the collection of statistics of manpower, production and distribution.

Manpower

It might seem strange, but in spite of the twentieth century being the age of manpower planning, it was not until 1945 that we had any reliable detailed information from the Ministry of Labour[2] about the size and distribution of the working population. Before 1939 there was highly reliable data relating to unemployment, but so far as the working population was concerned, all we had were estimates based on the Census of Population, the Census of Production and the Ministry of Labour's annual estimate. The war changed all that. It is impossible for a country to wage total war without detailed knowledge of the manpower available, and of industry's need for manpower. So the U.K. emerged from the war in 1945 with more information about manpower than ever before — but as you will appreciate it is not possible to rely on pre-war and post-war comparisons.

Just as the wartime planning necessitated the collection of manpower statistics, so post-war planning demanded, not only knowledge of the size

2. Now the Department of Employment and Productivity.

of the working population, but also its age and sex structure, its occupational skills, etc. Collection of such statistics was facilitated in 1948 by the passing of the National Insurance Act and the introduction of the Standard Industrial Classifications. The former required every gainfully employed person to register and in theory we should have obtained a complete census of all those in employment. Unfortunately, there are some omissions resulting from the failure of some individuals to register — particularly married women and the self-employed. The Standard Industrial Classification was designed to provide a standard pattern of classification by industry which could be used by all industrial statistics, and so facilitate comparison. Unfortunately from this point of view, there were major revisions in the classification in 1958 and 1968, and there have been many minor changes since. The effect of the major changes was to reclassify some industries as distributive and service industries which had previously been classed as manufacturing. You can see the effect this had in the following table, which classifies manpower according to two different Standard Industrial Classifications.

Change in the Composition of Employment (000's)
June 30th 1959

	[sic] 1948	[sic] 1958
Males		
Manufacturing	6,271	5,738
Distribution	1,578	1,689
Other	7,459	7,881
Females		
Manufacturing	2,898	2,739
Distribution	1,422	1,520
Other	3,569	3,630
Total	23,197	23,197

Source: Treasury Bulletin for Industry 1962.

The more recent changes have not been as drastic as this. They have involved largely the subdivision of industry into two or more component parts, as with the extracting in 1974 of 'Handtools and Implements' from the more general industry 'Cutlery Spoons Forks etc'.

	Thousand Employees		
	1972	1973	1974
Hand Tools and Implements	—	—	21
Cutlery Spoons Forks etc.	35	35	15

Unless you are very careful you can be left with the impression that there has been an unpheaval in certain industries at the time the classification was changed. However minor such changes may seem, however desirable they may be, they have one unfortunate effect — many of the series are not comparable over more than a few years.

One of the simplest ways of collecting data relating to employed persons was to use the quarterly exchange of insurance cards, since those who do hand in full insurance cards at the end of a quarter form a random sample of the working population. In fact, for many years, the annual figures published were based on cards handed in during the second quarter, supplemented by returns from employers having five or more workers. This gave a sample estimated to cover 75% of the working population. Now that the United Kingdom has abolished insurance cards, it seems that a major source, apart from the employers' annual return, will be found in the records of the Department of Health and Social Security or the Inland Revenue.

We would not wish to leave you with the impression that manpower statistics are concerned only with totals. Statistics are also produced relating to unemployment, unemployment rates, temporary lay-offs, days lost through strike action, average salaries and wages in the different industrial groups, holiday entitlement, etc. Moreover, most of the data is brought up to date monthly in the Department of Employment Gazette, and quarterly in the publications on Incomes, Prices, Employment and Production. Information today is far removed from the guesses and estimates of the 1930's.

Production
Without doubt one of the most important sources for data on industrial production is the Census of Production, the results of which are published by the Business Statistics Office. The full census is a massive document, but you will find a first-class summary of the relevant information in the Annual Abstract of Statistics. Although the expression 'census' is used, we must not imagine that it covers every firm in the United Kingdom. Firstly, there is a rather narrow definition of production. The census covers extractive, manufacturing and building industries. So important sectors of the economy such as agriculture and fisheries, distribution and the service industries are excluded. Secondly, even within this limited field, not every firm is included. The small firm is only required to give the nature of the trade and the size of its labour force. However, in some industries where a significant proportion of total output is produced by small firms, the full census requires such firms to make a simplified return.

In spite of its limited scope, the Census of Production provides us with a great deal of information. It shows us the way in which national industrial output is divided between different industries, and the study of this data over time brings out the trend and relative importance of individual industries. Central planning would be difficult without data like this, especially in respect of the distribution of labour and the allocation of capital resources. It is particularly important for the information it gives relating to changes in stocks of raw materials and finished goods, to capital expenditure on plant and machinery, and to building work undertaken. Without information such as this, national income statistics would be much less reliable, and the Index of Industrial Production subject to major errors.

Naturally, much of the information contained in the census report can be obtained elsewhere. The Department of Energy publishes much information on the coal industry and all forms of fuel production; The British Steel Corporation produces data on steel production; the building industry is covered by reports from the Department of the Environment. If you are interested in the details of a particular industry, say the tonnage of herrings landed at Hull, you would turn to the very comprehensive reports of the government departments, extracts from which appear in the Annual Abstract of Statistics. But the only place all this information is brought together is in the report on the Census of Production. Without this we would find it very difficult to estimate the total value of industrial production in the country.

In the United Kingdom, the first Census of Production was undertaken in 1907, and between the wars four more were taken. From 1948 it was intended to take a full census every year, but in fact the cost both to the state and to industry was heavy, and in the event a full census has been taken only every five years, while in the intervening years we have undertaken a sample survey, or a limited survey in which fewer questions are asked.

The census suffers from the normal defects of any effort to provide comprehensive and accurate information. It is expensive and it takes a considerable time to analyse and publish, and can only be undertaken at long intervals. To provide up-to-date information between censuses, the Department of Trade and Industry publish quarterly a series of sample statistics relating to different aspects of industry such as changes in stocks, capital expenditure and industrial buildings. The Department also sells statistical surveys of individual industries.

Distribution

Although Napoleon called England a nation of shopkeepers, it was not until 1951 that any official enquiry attempted to determine how many retail establishments there are in the United Kingdom, what kind of shops (multiples, independents) they were, what they sold or what was their turnover. The effort to obtain such information well illustrates the problems we face when we are attempting to take a census in a completely new field. Before any information could be obtained the government had to build up a sampling frame. This was done from May to October 1950, enumerators all over the country listing names and addresses of traders who fell within the scope of the census, 'as far as could be judged from the outside of the premises'. They were not allowed to enter the shop and talk to the owner or his employees, but were asked to distinguish between shops, stalls, depots and other kinds of premises. So thorough was this initial enumeration that the shops omitted must be negligible.

Having established a sampling frame the next step was to draft the questionnaire. This was the first time such a census had been undertaken, and a careful examination of objectives was necessary. It was decided that the census should provide information about

(a) the number and size of retail and wholesale outlets;
(b) the value of services provided by the distributive sector;
(c) the relative efficiency of distribution in different areas, i.e. the number of shops per head of the population.

Although the first census was a complete survey obtaining a 91% response rate to the questionnaire, a committee on the Census of Distribution and Production (the Vernon Smith Committee) recommended in 1954 that sample surveys be undertaken between the dates of the full census. It was decided to take a 12% sample and a most interesting thing about it was the way it was selected.

Firstly, every distributive outlet with a turnover in excess of £100,000 was to be included. This covered all the multiple stores, department stores and the majority of the co-operative societies. But once again there was no up-to-date sampling frame for the smaller independent retailers, so as in 1950 an enumeration had to be made in the relevant sample areas. These areas were chosen as follows:

(1) Certain areas existed in which it was believed that important changes had taken place since 1950, e.g. the new towns and central London. In these areas a complete enumeration of distributive outlets was undertaken and a 20% sample selected.

(2) In Greater London a sample of electoral wards stratified by size and distinguishing between mainly residential and mainly commercial was selected, and every shop in the selected wards was included in the survey.

(3) In the large towns (over 100,000 population) a sample of streets stratified by the number of shops there had been in 1950 was taken.

(4) Other towns were sampled by taking a cross-section of local authority areas stratified by population by sales in 1950, and by population changes since 1950.

(5) In rural districts a regional sample stratified by population density or population changes was selected.

The survey was undertaken by post with a very energetic follow-up. In the end the response rates were a remarkable 96% for larger traders, 89% for independent traders, and even 75% for street traders, hawkers, etc. It was estimated that the error was no more than 0.5% of the overall totals.

The 1961 Census of Distribution covered wholesale establishments which had been omitted from the sample survey, and to minimise cost only a 5% sample of establishments were given the full questionnaire. The remainder were asked only for details of employments and turnover together with some descriptive information. Thus we now obtain at five-year intervals details of the number, type and size of retail and wholesale firms, their turnover, stocks, capital expenditure, and hire purchase debt.

You may cast some doubt on the value of a census that takes nearly three years to publish its findings, and it is worth asking the reason for the delay. The first reason is the sheer magnitude of the operation — almost half a million retailers are involved, and it is no exaggeration to say that the returns take well over a year to come in. Even if a high proportion of firms are willing to co-operate the follow-up takes a considerable period of time. Time is absorbed too by a vast volume of correspondence from retailers asking questions about the completion of the forms.

Secondly, in the editing of the forms, many errors and omissions are found, and the correction of these errors involves both greater correspondence and delay. Finally, the analysis and printing of results, although computers are now used, is a tedious process.

Nevertheless, the importance of the analysis cannot be overestimated. Not only would National Income Statistics be unreliable without reasonably accurate figures of the contribution of the distributive trades to the Gross Domestic Product, but also the information is vital for the planning of, say, city shopping centres, or for the provision of shopping facilities in new towns. Moreover, the census returns form a new starting point for monthly and quarterly estimates which are vital in day-to-day planning.

National Income Statistics

All of the data that we have discussed in the last few paragraphs is still sectional data, no matter how widely the net is spread. It is concerned with only one aspect of the economy. What we must now do is turn our attention to the performance of the economy as a whole. This is seen only when we bring together the various parts of the economy in the statistics of National Income. We find these statistics in the book *National Income and Expenditure* published by Her Majesty's Stationery Office. This is not the place to go into details of what the expression 'National Income' means. However, we should note that all production is intended to satisfy consumer wants, and the more that is produced the greater can our consumption be. Thus what the National Income Statistics are measuring is basically the way our standard of living is changing.

Now any economist will tell you that National Income can be measured in three ways — by aggregating expenditure, by aggregating income, or by aggregating output. In theory, whichever method we use should given the same totals, but in practice there is always a difference which appears in the tables as 'residual error'. This has at times been as much as £700 million. Obviously we have not reached perfection!

To the extent that the data we have is accurate, overall planning is facilitated, but we must admit that despite a quarter of a century's experience in producing these figures, and despite many improvements that have taken place, several of the more important aggregates are little more than approximations. It is for this reason that each successive edition of *National Income and Expenditure* (called the 'Blue Book') amends the figures for the previous years.

There are differences in opinion as to what should be included in National Income calculations, and this makes international comparisons very difficult. For example in the peasant economies of the underdeveloped countries food grown by the farmers and eaten themselves is a very important part of the National Income and can be justifiably included. But in the United Kingdom such consumption is not included in the total of National Income. In fact in this country we include only such goods and services as are exchanged for money; and this cuts out a host of services which contribute to our welfare. If your wife or mother washes your shirt it does not affect the National Income statistics, but if you send it to a laundry the National Income is affected. Thus any change in social habits whereby people begin to pay for things they previously did for themselves will raise the National Income — even though the goods and services available have not increased.

One further point to be borne in mind is that the National Income is necessarily measured in money terms — and the value of money changes. Thus we may find that the National Income has risen by 10% since the last year, but if inflation is at the rate of 15% can we really say we are better off?

In spite of these criticisms, however, the annual Blue Book on National Income and Expenditure contains a mass of information vital to the planners in society and to the students of society.

Beginning with the National Income measured by the expenditure method, the Blue Book reconciles this figure with the National Income measured by the income of the various factors of production (wages, profits, interest and rent). This is far more than an academic exercise. The first table indicates the changing pattern of people's consumption. Here we can see for example, the changing expenditure on health, defence or education, the switch of expenditure to the service industries and the rise of the consumer durable industry. The second part of the table indicates the distribution of income in society; the proportion of income earned by the performance of work as compared with the proportion received from rents. Here we can trace the increasing share of the National Income going to the 'working class'. The third reconciliation indicates the Gross National Product by industry. Thus we can assess the contribution of each industrial group to the National Income. Other tables in the Blue Book take many of these points and in turn analyse them in much more detail.

This type of data is obtained from three main sources — the statistics assembled by the Inland Revenue, the Census of Production and Distribution, and the accounts of central and local government. But these are supplemented by a wide range of other sources. Unfortunately the coverage is far from complete, and in many cases the accuracy is impaired by the fact that the data used has been compiled for purposes other than National Income estimates.

The data we have discussed in this chapter is very important, but please remember that it is only a fraction of the data published each year. Much of this is a result of the day-to-day work of the government departments, some is an offspring of special surveys, and a great deal more is the result of the work of the statistical departments of banks, insurance companies, finance houses and the like. Overlooking and co-ordinating the work of the government departments is the Central Statistical Office producing the Annual Abstract of Statistics, the most important of all sources, and the starting point for a knowledge of what is available.

Exercises to Chapter 14

14.1 Describe where and in what form figures concerning population of the United Kingdom can be found. Say what use could be made of this information by an organisation of your own choice, e.g. an insurance company, a chain of supermarkets, a local authority or any other with which you are familiar.

14.2 Describe the statistical information which is available from the Census of Production.

14.3 Explain how and by whom the General Index of Retail Prices is prepared, where it is published and what purpose it serves.

14.4 Describe the construction of the Official Retail Prices Index used in the United Kingdom, and explain how it could be used by
    (i)   accountants
    (ii)   consumers
    (iii) trade unions

14.5 As a trade union official you have been given the task of submitting a well documented wage claim on behalf of your members.
    (a)   What kind of problems would you expect to encounter in making your case?
    (b)   What sources of economic statistics would you consult?
    (c)   What kind of information would you seek from those sources to substantiate your claim?

14.6 Describe the construction and use of two of the following Economic and Business Indicators.
    (a)   General Index of Retail Prices
    (b)   Index of Industrial Production
    (c)   Index Numbers of Wholesale Prices
    (d)   Unemployment and Employment Statistics

14.7 State with reasons FOUR important economic indicators which are used in charting the economy.

14.8 Discuss the statistical material available on National Income and Expenditure in the United Kingdom.

14.9 If your employer wished to gain information on the following suggest which publications would provide this:
(a) Wage rates and average earnings
(b) Retail and Wholesale prices
(c) Industrial production
(d) Regional and national unemployment.
Give examples of the purpose for which he may wish to use this information.

14.10 Outline the structure of the Family Expenditure Survey and explain its purpose. How reliable are its findings? Indicate some of the changes that have taken place over the last decade which affects the comparison with earlier years.

Answers to Exercises

1.1 115 pesetas = £0.78, 326 pesetas = £2.20, 428 pesetas = £2.89.

1.3 $m = \dfrac{74-45}{7-3} = \dfrac{32}{4} = 8$, $y = 8x + 21$

1.5 Maximum turning point when $x = -1$, minimum when $x = +1$.

1.7 When $x = 2$ 1   0   9 $a+1$ -5
         $y = 7$ 0 -5 112 a^2+6a 0

2.2 $mx^2 - c$ has a minimum value, $c - mx^2$ has a maximum value.

2.5 Average cost per mile = £0.00005x + $\dfrac{25.4}{400}$ + $\dfrac{180}{x}$

6000 miles per annum minimises cost per mile at 12.35p.

2.6 $\dfrac{ds}{dr} = 4\pi r - \dfrac{800}{r^2}$ $r = 3.993$ cm, $h = 7.989$ cm.

2.8 Total revenue = $80Q - 2Q^2$
T.R. maximised when $Q = 20$

2.10 Cost = $\dfrac{100000}{q} + \dfrac{0.001q}{4}$

Cost minimised when $q = 20{,}000$
order every two weeks.

3.1 a) $\begin{pmatrix} 3 & 5 & 2 \\ 7 & 1 & 0 \\ -2 & 7 & -1 \end{pmatrix}$ c) $\begin{pmatrix} 1 & -7 & -2 \\ -1 & 1 & 2 \\ 2 & 1 & -3 \end{pmatrix}$ e) $\begin{pmatrix} -2 & 12 & 5 \\ 5 & 21 & 6 \\ 20 & -6 & -6 \end{pmatrix}$

3.3 a) $\begin{pmatrix} 11 & 9 \\ 6 & 10 \\ 7 & 14 \end{pmatrix}$ c) not possible   e) not possible   g) $\begin{pmatrix} 93 \\ 66 \\ 84 \end{pmatrix}$

i) not possible.

3.5 $\begin{pmatrix} -1 & 2 & 0 \\ 2 & -2 & -1 \\ 0 & -½ & ½ \end{pmatrix}$

3.7 $\begin{pmatrix} a \\ b \\ c \end{pmatrix} = \begin{pmatrix} 13 \\ 22 \\ 39 \end{pmatrix}$ $\begin{pmatrix} a \\ b \\ c \\ d \end{pmatrix} = \begin{pmatrix} 1 \\ 2 \\ 3 \\ 4 \end{pmatrix}$

3.9 $T = \begin{pmatrix} 3/2 & 1 \\ 3/4 & 5/2 \end{pmatrix}$ $\begin{pmatrix} 10 \\ 45 \end{pmatrix} = \begin{pmatrix} 60 \\ 120 \end{pmatrix}$

3.11 $R = \begin{pmatrix} 240 & 0 \\ 0 & 300 \end{pmatrix}$, $P = \begin{pmatrix} 160 & 80 \\ 0 & 300 \end{pmatrix}$, $Q = \begin{pmatrix} 2/3 & 4/15 \\ 0 & 1 \end{pmatrix}$,

$S = \begin{pmatrix} 200 & 40 \\ 150 & 150 \end{pmatrix}$ $T = \begin{pmatrix} 5/6 & 2/15 \\ 5/8 & ½ \end{pmatrix}$

4.1 $S = 3250 (1.11)^7 = £6747.65$.

4.3 $(1 + r)^6 = \dfrac{5776}{3265}$, $r = 0.0997$ or 9.97%

4.5 $S = 9500 (1.12)^{12} + \dfrac{-800 (1.12)^{12} - (-800)}{0.12} = £17,705$

4.7 a) £23,380   b)i £3,108   b)ii £2,850.

4.9 £1,356

4.11 $125,000 \times 0.07 = £8750$ (annual income)
Market price $= 125,000(1.1)^{-4} + \dfrac{8750[1-(1.1)^{-4}]}{0.1} = £113,112$.

5.2 Payback period = 2.5 years
N.P.V. = £340.88,
N.P.V. at r = 0.28 = 1012.80, N.P.V. at 0.29 = 993.16,
I.R.R. = 28.65%.

5.4 P.V. = £11,666.40, so borrow and invest.

5.6 P.V. i) = £12000, ii) = £13210.45, iii) 13112.20, iv) £10222.10 so choose iv).

5.8 P.V. = £5294.60. IRR < 12%, so do not purchase.

5.10 a) N.P.V. = −£52.18, not viable
b) N.P.V. = £44.28, viable
c) N.P.V. = −£253.82, not viable.

6.9
Life (hours)	No. of bulbs
650 and under 660	2
660 –	5
670 –	6
680 –	14
690 –	26
700 –	18
710 –	13
720 –	10
730 –	3
740 –	3
	100

9.3 Median = 2.72, Mean = 2.48

9.5 Assume final group is £30 and under £40
assumed mean = £17.50, c = 5
$\Sigma f = 376$ (hundreds) $\Sigma fd = -224.5$ (hundreds)
Mean = £1,451.50

9.7 Assuming employment starts at 18, and retirement at 65, $\bar{x} = 43.16$ years.

9.10 Frequencies are 7, 16, 12, 11, 10, 4
$\Sigma f = 60$. If assumed mean = 549.5, and if c = 20
$\Sigma fd = 13$, $\bar{x} = 55383$ miles.

9.11 Factory A = 27.23 years, Factory B = 37.09 years.

9.14 Median = 84p, Mode = 76.9p
assumed mean of 75 and $C = 10$
$\Sigma f = 395$, $\Sigma fd = 453$, $\bar{x} = 86.47p$.

10.4 Trend 1347.5, 1382.5, 1440, 1532.5, 1608.75, 1658.75, 1715.5, 1800.0, 1878.75, 1926.25, 1997.5.

10.6 Seven Day moving average:
1137.1, 1145.7, 1148.6, 1148.6, 1148.6, 1150, 1148.6, 1155.7, 1168.6, 1174.3, 1172.9, 1168.6, 1171.4, 1172.9, 1181.4, 1202.9, 1207.1, 1205.7.

10.8 Trend
2.55, 2.725, 2.95, 3.21, 3.475, 3.64, 3.675, 3.61, 3.41, 3.16, 2.9, 2.61, 2.425, 2.36.

11.2 Trend: 14.875, 16.0, 16.75, 16.875, 17.5, 18.625, 19.375, 19.5, 19.625, 20.75, 22.5, 23.625.
Actual/Trend: 1.0756, 0.5, 0.8358, 1.7185, 0.9142, 0.4832, 0.929, 1.7435, 0.8662, 0.4337, 1.7777.

11.4 Trend 350.75, 350.625, 347.75, 341.625, 334.125, 325.875, 313.125, 299.125, 290.125.
Deviations -0.75, 44.375, -14.75, -21.625, 1.875, 34.125, -5.125, -20.125, -15.125.
Seasonal Variation: -11, -22, $-5\frac{1}{2}$, $+38\frac{1}{2}$.

11.6 Trend 505.625, 515.875, 515.25, 510, 507.625, 507.625, 514.625, 523.875.
Deviations -39.625, 10.125, 25.75, 31, -67.625, 2.375, 23.375, 20.125
Seasonal Variation $+24$, $+25$, -54.5, $+5.5$
Deseasonalised Series 456, 495, 520.5, 520.5, 517, 516, 494.5, 504.5, 514, 519, 547.5, 525.5.

11.8 Trend 180.125, 183.75, 205.125, 238, 251.625, 261.875, 259.25, 252.25, 254.125, 249, 245.625, 245.625
Deviations -46.125, 29.25, -18.125, -40, 59.375, 37.125, -49.25, 4.75, -23.125, 74, -44.625, -20.625
Seasonal Variation -34, -16, 0, $+50$.

11.10 Trend 13.125, 13.5, 13.875, 14.375, 15.125, 15.75, 16.375, 16.875
Actual/Trend 1.219, 0.6666, 1.009, 1.113, 1.1239, 0.7619, 1.0381, 1.0666
Seasonal Variation 1.09, 1.1718, 0.7144, 1.0238
Deasonalised Data 13, 13, 14, 13, 14, 15, 14.5, 17, 17, 17, 17, 18

11.13 With $\alpha = 0.5$, forecasts are 13, 13.5, 14.75, 11.88, 12.94, 14.47, 15.73, 13.87, 15.43, 16.71, 18.35, 15.68.

12.2 Laspeyre's Price Index for 19-4 = $\dfrac{56.24}{60.42} \times 100 = 93.1$

Laspeyre's Price Index for 19-5 = $\dfrac{59.14}{60.42} \times 100 = 97.9$

12.4 a) Paasche Price Index for 19-4 = $\dfrac{2460}{2118} \times 100 = 116.1$

b)
	Weight
Ingots	430
Bars	355
Strip	415
	1000

12.6 Price Index

      Laspeyre's $\quad \dfrac{49.75}{37.5} \times 100 = 132.7$

      Paasche's $\quad \dfrac{49.0}{37.0} \times 100 = 132.4$

Volume Index

      Laspeyre's $\quad \dfrac{37.0}{37.5} \times 100 = 98.7$

      Paasche's $\quad \dfrac{49.0}{49.75} \times 100 = 98.5$

12.8 Laspeyre's Volume Index

      for 19-1 $\quad \dfrac{533}{522} \times 100 = 102.1$

      for 19-2 $\quad \dfrac{469}{522} \times 100 = 89.8$

Paasches Price Index

      for 19-1 $\quad \dfrac{430}{533} \times 100 = 80.7$

      for 19-2 $\quad \dfrac{351}{469} \times 100 = 74.8$

12.10

Year	Profits at 1973 Prices	Index of 'real' profits, 1973 = 100
1970	1375	97.4
1971	1408	99.7
1972	1420	100.6
1973	1412	100.0
1974	1433	101.5
1975	1493	105.7
1976	1480	104.8
1977	1535	108.7
1978	1563	110.7
1979	1607	113.8

13.1 Range = 67, Mean = 48, Median = 48, L.Q. = 36, U.Q. = 63, Q.D. = 13.5, Mean Deviation = 14.08.

13.4 With assumed mean of 75, and unit of 10,
$\Sigma f = 65$, $\Sigma fd = 31$, $\Sigma fd^2 = 173$
$\bar{x} = £79.77$
$\sigma = £15.60$.

13.5 Assuming last group is 90–94, assumed mean = 77.5, unit of 5, $\Sigma f = 2650$, $\Sigma fd = -2420$, $\Sigma fd^2 = 5680$
$\bar{x} = 72.93$ years
$\sigma = 5.72$ years.

13.6 $\bar{x} = 4.375$ tons, $\sigma = 2.94$ tons

13.9

	A	B
Mean	£36.60	£26.00
S.D.	£2.82	£2.39
C of V	7.70%	9.19%

Compounding Tables for R = .01 to R = .07

	1.01	1.02	1.03	1.04	1.05	1.06	1.07
1	1.0100	1.0200	1.0300	1.0400	1.0500	1.0600	1.0700
2	1.0201	1.0404	1.0609	1.0816	1.1025	1.1236	1.1449
3	1.0303	1.0612	1.0927	1.1249	1.1576	1.1910	1.2250
4	1.0406	1.0824	1.1255	1.1699	1.2155	1.2625	1.3108
5	1.0510	1.1041	1.1593	1.2167	1.2763	1.3382	1.4026
6	1.0615	1.1262	1.1941	1.2653	1.3401	1.4185	1.5007
7	1.0721	1.1487	1.2299	1.3159	1.4071	1.5036	1.6058
8	1.0829	1.1717	1.2668	1.3686	1.4775	1.5938	1.7182
9	1.0937	1.1951	1.3048	1.4233	1.5513	1.6895	1.8385
10	1.1046	1.2190	1.3439	1.4802	1.6289	1.7908	1.9672
11	1.1157	1.2434	1.3842	1.5395	1.7103	1.8983	2.1049
12	1.1268	1.2682	1.4258	1.6010	1.7959	2.0122	2.2522
13	1.1381	1.2936	1.4685	1.6651	1.8856	2.1329	2.4098
14	1.1495	1.3195	1.5126	1.7317	1.9799	2.2609	2.5785
15	1.1610	1.3459	1.5580	1.8009	2.0789	2.3966	2.7590
16	1.1726	1.3728	1.6047	1.8730	2.1829	2.5404	2.9522
17	1.1843	1.4002	1.6528	1.9479	2.2920	2.6928	3.1588
18	1.1961	1.4282	1.7024	2.0258	2.4066	2.8543	3.3799
19	1.2081	1.4568	1.7535	2.1068	2.5270	3.0256	3.6165
20	1.2202	1.4859	1.8061	2.1911	2.6533	3.2071	3.8697

Compounding Tables for R = .08 to R = .14

	1.08	1.09	1.10	1.11	1.12	1.13	1.14
1	1.0800	1.0900	1.1000	1.1100	1.1200	1.1300	1.1400
2	1.1664	1.1881	1.2100	1.2321	1.2544	1.2769	1.2996
3	1.2597	1.2950	1.3310	1.3676	1.4049	1.4429	1.4815
4	1.3605	1.4116	1.4641	1.5181	1.5735	1.6305	1.6890
5	1.4693	1.5386	1.6105	1.6851	1.7623	1.8424	1.9254
6	1.5869	1.6771	1.7716	1.8704	1.9738	2.0820	2.1950
7	1.7138	1.8280	1.9487	2.0762	2.2107	2.3526	2.5023
8	1.8509	1.9926	2.1436	2.3045	2.4760	2.6584	2.8526
9	1.9990	2.1719	2.3579	2.5580	2.7731	3.0040	3.2519
10	2.1589	2.3674	2.5937	2.8394	3.1058	3.3946	3.7072
11	2.3316	2.5804	2.8531	3.1518	3.4785	3.8359	4.2262
12	2.5182	2.8127	3.1384	3.4985	3.8960	4.3345	4.8179
13	2.7196	3.0658	3.4523	3.8833	4.3635	4.8980	5.4924
14	2.9372	3.3417	3.7975	4.3104	4.8871	5.5348	6.2613
15	3.1722	3.6425	4.1772	4.7846	5.4736	6.2543	7.1379
16	3.4259	3.9703	4.5950	5.3109	6.1304	7.0673	8.1372
17	3.7000	4.3276	5.0545	5.8951	6.8660	7.9861	9.2765
18	3.9960	4.7171	5.5599	6.5436	7.6900	9.0243	10.5752
19	4.3157	5.1417	6.1159	7.2633	8.6128	10.1974	12.0557
20	4.6610	5.6044	6.7275	8.0623	9.6463	11.5231	13.7435

Compounding Tables for R = .15 to R = .21

	1.15	1.16	1.17	1.18	1.19	1.20	1.21
1	1.1500	1.1600	1.1700	1.1800	1.1900	1.2000	1.2100
2	1.3225	1.3456	1.3689	1.3924	1.4161	1.4400	1.4641
3	1.5209	1.5609	1.6016	1.6430	1.6852	1.7280	1.7716
4	1.7490	1.8106	1.8739	1.9388	2.0053	2.0736	2.1436
5	2.0114	2.1003	2.1924	2.2878	2.3864	2.4883	2.5937
6	2.3131	2.4364	2.5652	2.6996	2.8398	2.9860	3.1384
7	2.6600	2.8262	3.0012	3.1855	3.3793	3.5832	3.7975
8	3.0590	3.2784	3.5115	3.7589	4.0214	4.2998	4.5950
9	3.5179	3.8030	4.1084	4.4355	4.7854	5.1598	5.5599
10	4.0456	4.4114	4.8068	5.2338	5.6947	6.1917	6.7275
11	4.6524	5.1173	5.6240	6.1759	6.7767	7.4301	8.1403
12	5.3503	5.9360	6.5801	7.2876	8.0642	8.9161	9.8497
13	6.1528	6.8858	7.6987	8.5994	9.5964	10.6993	11.9182
14	7.0757	7.9875	9.0075	10.1472	11.4198	12.8392	14.4210
15	8.1371	9.2655	10.5387	11.9737	13.5895	15.4070	17.4494
16	9.3576	10.7480	12.3303	14.1290	16.1715	18.4884	21.1138
17	10.7613	12.4677	14.4265	16.6722	19.2441	22.1861	25.5477
18	12.3755	14.4625	16.8790	19.6733	22.9005	26.6233	30.9127
19	14.2318	16.7765	19.7484	23.2144	27.2516	31.9480	37.4043
20	16.3665	19.4608	23.1056	27.3930	32.4294	38.3376	45.2593

Compounding Tables for R = .22 to R = .28

	1.22	1.23	1.24	1.25	1.26	1.27	1.28
1	1.2200	1.2300	1.2400	1.2500	1.2600	1.2700	1.2800
2	1.4884	1.5129	1.5376	1.5625	1.5876	1.6129	1.6384
3	1.8158	1.8609	1.9066	1.9531	2.0004	2.0484	2.0972
4	2.2153	2.2889	2.3642	2.4414	2.5205	2.6014	2.6844
5	2.7027	2.8153	2.9316	3.0518	3.1758	3.3038	3.4360
6	3.2973	3.4628	3.6352	3.8147	4.0015	4.1959	4.3980
7	4.0227	4.2593	4.5077	4.7684	5.0419	5.3288	5.6295
8	4.9077	5.2389	5.5895	5.9605	6.3528	6.7675	7.2058
9	5.9874	6.4439	6.9310	7.4506	8.0045	8.5948	9.2234
10	7.3046	7.9259	8.5944	9.3132	10.0857	10.9153	11.8059
11	8.9117	9.7489	10.6571	11.6415	12.7080	13.8625	15.1116
12	10.8722	11.9912	13.2148	14.5519	16.0120	17.6053	19.3428
13	13.2641	14.7491	16.3863	18.1899	20.1752	22.3588	24.7588
14	16.1822	18.1414	20.3191	22.7374	25.4207	28.3957	31.6913
15	19.7423	22.3140	25.1956	28.4217	32.0301	36.0625	40.5648
16	24.0856	27.4462	31.2426	35.5271	40.3579	45.7994	51.9230
17	29.3844	33.7588	38.7408	44.4089	50.8510	58.1652	66.4614
18	35.8490	41.5233	48.0386	55.5112	64.0722	73.8698	85.0706
19	43.7358	51.0737	59.5679	69.3889	80.7310	93.8147	108.8904
20	53.3576	62.8206	73.8642	86.7362	101.7211	119.1446	139.3797

Compounding Tables for R = .29 to R = .35

	1.29	1.30	1.31	1.32	1.33	1.34	1.35
1	1.2900	1.3000	1.3100	1.3200	1.3300	1.3400	1.3500
2	1.6641	1.6900	1.7161	1.7424	1.7689	1.7956	1.8225
3	2.1467	2.1970	2.2481	2.3000	2.3526	2.4061	2.4604
4	2.7692	2.8561	2.9450	3.0360	3.1290	3.2242	3.3215
5	3.5723	3.7129	3.8579	4.0075	4.1616	4.3204	4.4840
6	4.6083	4.8268	5.0539	5.2899	5.5349	5.7893	6.0534
7	5.9447	6.2749	6.6206	6.9826	7.3614	7.7577	8.1722
8	7.6686	8.1573	8.6730	9.2170	9.7907	10.3953	11.0324
9	9.8925	10.6045	11.3617	12.1665	13.0216	13.9297	14.8937
10	12.7614	13.7858	14.8838	16.0598	17.3187	18.6659	20.1066
11	16.4622	17.9216	19.4977	21.1989	23.0339	25.0123	27.1439
12	21.2362	23.2981	25.5420	27.9825	30.6351	33.5164	36.6442
13	27.3947	30.2875	33.4601	36.9370	40.7447	44.9120	49.4697
14	35.3391	39.3738	43.8327	48.7568	54.1905	60.1821	66.7841
15	45.5875	51.1859	57.4208	64.3590	72.0733	80.6440	90.1585
16	58.8079	66.5417	75.2213	84.9538	95.8575	108.0629	121.7139
17	75.8621	86.5042	98.5399	112.1390	127.4905	144.8043	164.3138
18	97.8622	112.4554	129.0872	148.0235	169.5624	194.0378	221.8236
19	126.2422	146.1920	169.1043	195.3911	225.5180	260.0107	299.4619
20	162.8524	190.0496	221.5266	257.9162	299.9389	348.4143	404.2736

Discounting Tables for R = .01 to R = .07

	1.01	1.02	1.03	1.04	1.05	1.06	1.07
1	.9901	.9804	.9709	.9615	.9524	.9434	.9346
2	.9803	.9612	.9426	.9246	.9070	.8900	.8734
3	.9706	.9423	.9151	.8890	.8638	.8396	.8163
4	.9610	.9238	.8885	.8548	.8227	.7921	.7629
5	.9515	.9057	.8626	.8219	.7835	.7473	.7130
6	.9420	.8880	.8375	.7903	.7462	.7050	.6663
7	.9327	.8706	.8131	.7599	.7107	.6651	.6227
8	.9235	.8535	.7894	.7307	.6768	.6274	.5820
9	.9143	.8368	.7664	.7026	.6446	.5919	.5439
10	.9053	.8203	.7441	.6756	.6139	.5584	.5083
11	.8963	.8043	.7224	.6496	.5847	.5268	.4751
12	.8874	.7885	.7014	.6246	.5568	.4970	.4440
13	.8787	.7730	.6810	.6006	.5303	.4688	.4150
14	.8700	.7579	.6611	.5775	.5051	.4423	.3878
15	.8613	.7430	.6419	.5553	.4810	.4173	.3624
16	.8528	.7284	.6232	.5339	.4581	.3936	.3387
17	.8444	.7142	.6050	.5134	.4363	.3714	.3166
18	.8360	.7002	.5874	.4936	.4155	.3503	.2959
19	.8277	.6864	.5703	.4746	.3957	.3305	.2765
20	.8195	.6730	.5537	.4564	.3769	.3118	.2584

Discounting Tables for R = .08 to R = .14

	1.08	1.09	1.10	1.11	1.12	1.13	1.14
1	.9259	.9174	.9091	.9009	.8929	.8850	.8772
2	.8573	.8417	.8264	.8116	.7972	.7831	.7695
3	.7938	.7722	.7513	.7312	.7118	.6931	.6750
4	.7350	.7084	.6830	.6587	.6355	.6133	.5921
5	.6806	.6499	.6209	.5935	.5674	.5428	.5194
6	.6302	.5963	.5645	.5346	.5066	.4803	.4556
7	.5835	.5470	.5132	.4817	.4523	.4251	.3996
8	.5403	.5019	.4665	.4339	.4039	.3762	.3506
9	.5002	.4604	.4241	.3909	.3606	.3329	.3075
10	.4632	.4224	.3855	.3522	.3220	.2946	.2697
11	.4289	.3875	.3505	.3173	.2875	.2607	.2366
12	.3971	.3555	.3186	.2858	.2567	.2307	.2076
13	.3677	.3262	.2897	.2575	.2292	.2042	.1821
14	.3405	.2992	.2633	.2320	.2046	.1807	.1597
15	.3152	.2745	.2394	.2090	.1827	.1599	.1401
16	.2919	.2519	.2176	.1883	.1631	.1415	.1229
17	.2703	.2311	.1978	.1696	.1456	.1252	.1078
18	.2502	.2120	.1799	.1528	.1300	.1108	.0946
19	.2317	.1945	.1635	.1377	.1161	.0981	.0829
20	.2145	.1784	.1486	.1240	.1037	.0868	.0728

Discounting Tables for R = .15 to R = .21

	1.15	1.16	1.17	1.18	1.19	1.20	1.21
1	.8696	.8621	.8547	.8475	.8403	.8333	.8264
2	.7561	.7432	.7305	.7182	.7062	.6944	.6830
3	.6575	.6407	.6244	.6086	.5934	.5787	.5645
4	.5718	.5523	.5337	.5158	.4987	.4823	.4665
5	.4972	.4761	.4561	.4371	.4190	.4019	.3855
6	.4323	.4104	.3898	.3704	.3521	.3349	.3186
7	.3759	.3538	.3332	.3139	.2959	.2791	.2633
8	.3269	.3050	.2848	.2660	.2487	.2326	.2176
9	.2843	.2630	.2434	.2255	.2090	.1938	.1799
10	.2472	.2267	.2080	.1911	.1756	.1615	.1486
11	.2149	.1954	.1778	.1619	.1476	.1346	.1228
12	.1869	.1685	.1520	.1372	.1240	.1122	.1015
13	.1625	.1452	.1299	.1163	.1042	.0935	.0839
14	.1413	.1252	.1110	.0985	.0876	.0779	.0693
15	.1229	.1079	.0949	.0835	.0736	.0649	.0573
16	.1069	.0930	.0811	.0708	.0618	.0541	.0474
17	.0929	.0802	.0693	.0600	.0520	.0451	.0391
18	.0808	.0691	.0592	.0508	.0437	.0376	.0323
19	.0703	.0596	.0506	.0431	.0367	.0313	.0267
20	.0611	.0514	.0433	.0365	.0308	.0261	.0221

Discounting Tables for R = .22 to R = .28

	1.22	1.23	1.24	1.25	1.26	1.27	1.28
1	.8197	.8130	.8065	.8000	.7937	.7874	.7812
2	.6719	.6610	.6504	.6400	.6299	.6200	.6104
3	.5507	.5374	.5245	.5120	.4999	.4882	.4768
4	.4514	.4369	.4230	.4096	.3968	.3844	.3725
5	.3700	.3552	.3411	.3277	.3149	.3027	.2910
6	.3033	.2888	.2751	.2621	.2499	.2383	.2274
7	.2486	.2348	.2218	.2097	.1983	.1877	.1776
8	.2038	.1909	.1789	.1678	.1574	.1478	.1388
9	.1670	.1552	.1443	.1342	.1249	.1164	.1084
10	.1369	.1262	.1164	.1074	.0992	.0916	.0847
11	.1122	.1026	.0938	.0859	.0787	.0721	.0662
12	.0920	.0834	.0757	.0687	.0625	.0568	.0517
13	.0754	.0678	.0610	.0550	.0496	.0447	.0404
14	.0618	.0551	.0492	.0440	.0393	.0352	.0316
15	.0507	.0448	.0397	.0352	.0312	.0277	.0247
16	.0415	.0364	.0320	.0281	.0248	.0218	.0193
17	.0340	.0296	.0258	.0225	.0197	.0172	.0150
18	.0279	.0241	.0208	.0180	.0156	.0135	.0118
19	.0229	.0196	.0168	.0144	.0124	.0107	.0092
20	.0187	.0159	.0135	.0115	.0098	.0084	.0072

Discounting Tables for R = .29 to R = .35

	1.29	1.30	1.31	1.32	1.33	1.34	1.35
1	.7752	.7692	.7634	.7576	.7519	.7463	.7407
2	.6009	.5917	.5827	.5739	.5653	.5569	.5487
3	.4658	.4552	.4448	.4348	.4251	.4156	.4064
4	.3611	.3501	.3396	.3294	.3196	.3102	.3011
5	.2799	.2693	.2592	.2495	.2403	.2315	.2230
6	.2170	.2072	.1979	.1890	.1807	.1727	.1652
7	.1682	.1594	.1510	.1432	.1358	.1289	.1224
8	.1304	.1226	.1153	.1085	.1021	.0962	.0906
9	.1011	.0943	.0880	.0822	.0768	.0718	.0671
10	.0784	.0725	.0672	.0623	.0577	.0536	.0497
11	.0607	.0558	.0513	.0472	.0434	.0400	.0368
12	.0471	.0429	.0392	.0357	.0326	.0298	.0273
13	.0365	.0330	.0299	.0271	.0245	.0223	.0202
14	.0283	.0254	.0228	.0205	.0185	.0166	.0150
15	.0219	.0195	.0174	.0155	.0139	.0124	.0111
16	.0170	.0150	.0133	.0118	.0104	.0093	.0082
17	.0132	.0116	.0101	.0089	.0078	.0069	.0061
18	.0102	.0089	.0077	.0068	.0059	.0052	.0045
19	.0079	.0068	.0059	.0051	.0044	.0038	.0033
20	.0061	.0053	.0045	.0039	.0033	.0029	.0025

Index

Adaptive Forecasting 163-165
Annual Abstract of Statistics 197
Annual Average Rate of Return 55-57
Annuity 49-50
Arithmetic Mean 124-132
Associative 28
-value 165

Bar Charts 85-88
  Component 87-88
  Compound 87-88
Base Period 178-179

Central Tendency Chapter 9
  Arithmetic Mean 124-132
  Median 132-136
  Mode 136-138
Census of distribution 206-8
  of population 198-200
  of production 205-6
Chain Base Index 178-179
Chard 9
Circular Diagram 88-9
Class Interval 76
Commutative 26
Compounding and discounting
  Chapter 4
Compound Interest 38-42
Constant 1
Co-ordinates 2
Cumulative Frequency Diagram 91-92

Data
  Continuous 76
  Discrete 76
  Primary 70
  Secondary 70
Department of Employment Gazette
  205
Derivative 11
  First 16
  Second 16
Determinant 31
Deviation 125
Diagrams Ch. 7, Ch. 8

Differentiation Chapter 2
  Fractional Indices 11
  From first principles 9-11
  Negative Indices 11
  Products 12-13
  Quotients 13-14
  Rules for 11
Dispersion Ch. 13
  Coefficient of variation 191-192
  interquartile range 185-6
  mean absolute deviation 187-88
  quartile coefficient of variation 192
  quartile deviation 186
  range 184-185
  relative dispersion 191-2
  standard deviation 189-191
  variance 188

Expenditure Index 173-4
Exponential Smoothing 165-169

Family Expenditure Survey 176,
  201-202
Fisher's Index 179-180
Frequency
  Cumulative 92
  Curve 92
  Distribution 74-75
  Polygon 91
Functions 1
  Linear 5

Gradients 4-7, 9-11

Histogram 89-91

Ideagraph 115
Ideal Index 179-180
Identification
  of maxima 15-18
  of minima 15-18
Identity Matrix 26-31
Index Numbers Ch. 12
  base period 178-179
  chain base 178-179
  defined 173
  deflating a series 180

expenditure 173-174
Fisher's ideal 179-180
Laspeyre's 174-176
Paasche's 176-177
Price relative 175
simple aggregative 174-175
volume index 177
weighting 175
Index of Retail Prices 173, 200-206
Interquartile range 185-6
Internal Rate of Return 61-63
Inverse 27-31
Investment Evaluation Chapter 5

Laspeyre's Index 174-176
Linear functions 5
   and time series 159-161
Loan Repayment 45-47
Lorenz Curve 109-113

Manpower statistics 203-205
Matrix Chapter 3
   Addition 22-23
   Multiplication 23-28
   Identity 26-31
   Inversion 29-31
   Subtraction 22-33
Maxima 15-18
Mean, Arithmetic 124-132
Median 132-136
Mid-Point 76-77
Minima 15-18
Mode 136-138
Moving Average 146-149

National Income Statistics 208-210
Net Present Value 58-61

Ogive 91-92

Paasche's Index 176-177
Parabola 6
Partitioned Matrix 29-31
Payback period 53
Percentage Strata graph 97-98
Pictogram 115
Pie Diagram 88-89
Population 74
Present Value 47-50
Price Relative 175
Primary Data 70

Questionnaire 69-70

Range 77, 184-185
Ratio Scale graph 103-108
Relative Dispersion 191-2
Residuals 155-156
Retail Price Index 173, 200-206
Row Transformations 28

Sample 74
Scatter Diagram 114-115
Seasonal Variation 145, 152-155,
   159-160, 162-163
   elimination of 154-155, 162-163
second derivative 16-18
Secondary Data 70
Semi-logarithmic graph 103-108
Sensitivity 163
Set 20
Simple Aggregative Index 174-175
Simple Interest 38-39
Simultaneous Equations 32-34
Sinking Fund 44-45
Stability 164
Standard deviation 189-191
Standard Industrial Classification 204
Strata graph 97-98

Tabulation 71-73
   principles of 73
Time series 79, Ch. 10, Ch. 11
   Additive Model 159-163
   Forecasting 157-159
   Multiplicative Model 161-163
   Residuals 155-156
   Seasonal variation 152-155, 159-160,
      162-163
   Trend 136, 148
Trend — calculation of 146-149
   centreing 149
   linear 159-161
Trust funds 45

Unit Matrix 26-31

Variables 1, 70
   dependent 1
   independent 1
Variance 188
Volume Index 177
Vector 20

Z-chart 113-114